Tragedy at Chu Lai

Tragedy at Chu Lai

*Reconstructing a Deadly Grenade
Accident in a U.S. Army Classroom
in Vietnam, July 10, 1969*

DAVID VENDITTA

McFarland & Company, Inc., Publishers
Jefferson, North Carolina

LIBRARY OF CONGRESS CATALOGUING-IN-PUBLICATION DATA

Names: Venditta, David, 1954– author.
Title: Tragedy at Chu Lai : reconstructing a deadly grenade
    accident in a U.S. Army classroom in Vietnam, July 10, 1969 /
    David Venditta.
Other titles: Reconstructing a deadly grenade accident in a U.S.
    Army classroom in Vietnam, July 10, 1969
Description: Jefferson, North Carolina : McFarland & Company,
    Inc., Publishers, 2016 | Includes bibliographical references and
    index.
Identifiers: LCCN 2016022316 | ISBN 9781476664316 (softcover :
    acid free paper) ∞
Subjects: LCSH: Venditti, Nicky, 1948–1969. | United States. Army.
    Americal Division—Biography. | Vietnam War, 1961–1975—
    Vietnam—Chu Lai. | Military education—Accidents—United
    States. | Grenades—Accidents—Vietnam. | Helicopter pilots—
    United States—Biography. | Helicopter pilots—Vietnam—Biog-
    raphy. | United States. Army—Warrant officers—Biography. |
    Vietnam War, 1961–1975—Biography. | Malvern, Pa.—
    Biography.
Classification: LCC DS558.8 .V45 2016 | DDC 959.704/342—dc23
LC record available at https://lccn.loc.gov/2016022316

BRITISH LIBRARY CATALOGUING DATA ARE AVAILABLE

ISBN (print) 978-1-4766-6431-6
ISBN (ebook) 978-1-4766-2438-9

Front cover: U.S. soldier and American flag © 2016 Marko Misic/iStock

Printed in the United States of America

McFarland & Company, Inc., Publishers
    Box 611, Jefferson, North Carolina 28640
        www.mcfarlandpub.com

To the memories of
Nicky Venditti,
Billy Vachon and Tim Williams
and to their families

Table of Contents

Acknowledgments

Over the course of two decades, many people helped me with this book. I thank my extended family, my cousin Nicky Venditti's friends, the families of Billy Vachon and Tim Williams, and the Vietnam veterans who spoke with me.

Dick Bielen of U.S. Locator Service was indispensable in tracking down veterans I needed to reach. Investigative reporter Don Ray offered his time and expertise, and urged me on. Military author George Lepre shared his research know-how and command of the Army bureaucracy. Dick Boylan of the National Archives at College Park, Maryland, sleuthed for records and guided me around the place as part of the hunt.

Vietnam helicopter pilot and instructor Kevin Bagley answered my questions about the aircraft and flight school. Americal Division historian Les Hines provided scores of documents, photos, videos, maps and news clippings on the division and the Chu Lai base. Americal veteran Bob Short conveyed the U.S. soldier's experience in Vietnam and reviewed a draft. My family doctor, Joe Habig, translated Nicky's clinical records into layman's terms.

Longtime friend Steve Young discussed the story with me at our favorite diners and read the manuscript like a prosecutor. Early on, my newspaper co-workers Christine Schiavo, John P. Martin and Frank Warner helped me see what I had to do. Victoria Adam of Bijou Travel/ Incredible Journeys set up my Vietnam trip with tour operator Absolute Asia. When I arrived at Ho Chi Minh City, anxious about my mission, my friend Le Quoc Vinh put my concerns to rest.

Editors played a huge role. Polly Kummel recommended a key change in chapter order. In 2009, when I was struggling with the project, Ellen Roberts revived my spirits with her enthusiasm for the story.

Elizabeth Zack suggested broadening its scope. A former co-worker turned lawyer, Becky Jones, critiqued my first effort, a raw 146 pages.

My wife, Mary, has said the best move I made was hiring Ardith Hilliard. An astute mentor, Ardie convinced me that the story could not merely be reported objectively, but had to be a first-person account of my journey of discovery. She helped me see the arc of the narrative, offered praise when she liked what she read, prodded me when I needed to do better. Always, she encouraged me to look deep inside myself and give voice to my feelings and long-ago memories.

Ardie had come to *The Morning Call* in Allentown, Pennsylvania, from her job as an editor at the *Los Angeles Times,* where she worked on teams that won three Pulitzer Prizes. While at the Allentown newspaper, where I was an editor, she served on Pulitzer selection committees for two years. She was a retired executive editor in 2011 when we started our collaboration.

My stepdaughters, Rachel and Teresa Steigerwalt, endured my absences and distractions. Mary was my companion throughout, sharing my triumphs and disappointments, demonstrating courage in tense moments and helping me sort out the story. She put up with the demands of a project that intruded on countless days and nights we might have spent together. Her love sustained me.

Preface

During the twenty-one years I've worked on my cousin Nicky Venditti's story and his inexplicable death in Vietnam, I have held tightly to the belief that writing about someone who is gone brings that person back to life. I wanted desperately to create a permanent record for family and friends so Nicky and his sacrifice would abide for generations. What I couldn't have imagined at the start was how much time and effort bringing Nicky back would take, that it would alternate between immensely satisfying and profoundly exasperating, that it would consume most of my middle age and that it ultimately would lead to my writing a book.

The extent of my commitment first struck me on a wintry Sunday in February 1996. I was visiting Mary Anne Wallace, one of Nicky's close friends and the first person outside my extended family I spoke with about him. After we had chatted a few hours in her home outside West Chester, Pennsylvania, and were saying goodbye, she said of my plan to tell Nicky's story: "You know this will take you a couple of years."

Oh no, I thought, instantly shaken. It seemed like an impossibly long time, but she was probably right. I wondered what I was getting myself into, how I could pull this off.

Back home that afternoon, I had an hour or so and decided to make a call. I was pumped up by what Mary Anne had told me about Nicky and their friendship, and eager to make more progress. If I could just find Tony Viall, one of my cousin's friends and a survivor of the explosion that left Nicky mortally wounded, surely he could tell me what happened. Tony's name and hometown in Georgia were on a roster of pilots in Nicky's Army paperwork, so I called a Viall listed in the town. Tony's mother answered. She gave me his phone number and where he lived.

I was on my way. Just by tugging on one string a little, almost casually, I had steered myself onto a true path—a reward I would see again and again over the years as I pulled whatever loose strings came within my grasp.

As you are reading, you will come across some dialogue, including quotes from Nicky. These quotes reflect how those involved—such as whoever was with Nicky at the time—remember the conversations. Where there are people quoted or paraphrased and not named, it is because I do not have their consent. I have interviewed more than 130 people and sought help from dozens of others, and I am grateful to all who put their trust in me. Almost every person agreed to be identified and quoted, but I have taken care to keep the privacy of those I wish to protect or who have their reasons not to be named.

The spelling of Nicky's last name differs from mine. When my grandfather arrived at Ellis Island from Italy, his name was altered from Venditti to Venditta. Nicky's dad and another of Grandpop's sons reclaimed the original name. My dad and other brothers did not.

Introduction:
A Vietnam Odyssey,
May 18, 1998

The driver beeped his horn in staccato bursts, our car vying with bicycles and motorbikes for a path on the barely two-lane Highway 1 outside Da Nang. As the biggest city on Vietnam's central coast disappeared behind us, the bikes tailed off the pavement like geckos I'd seen darting up and down the walls of apartments and cafés.

My guide, Dinh Mien, sat up front with driver Tran Mroc Manh, and I rode alone in the back. Above us the sun had scoured the sky into a faint blue, leaving only streaks and puffs of white clouds that hung as if painted there. Heat bore down heavily though it was barely eight in the morning. Still, with fifty-six miles to go and Manh driving at a leisurely forty miles an hour, we had time to stay cool in our air-conditioned black Mazda sedan.

Our destination to the south was a chunk of coast called Chu Lai, a restricted area under the control of Vietnam's military. During the Vietnam War, it was one of the largest U.S. bases in South Vietnam.

It was where my cousin Nicky Venditti died.

He had arrived at Chu Lai in the summer of 1969 as a twenty-year-old Army helicopter pilot. A week and a half later, he was dead.

Although we shared a close, warm, sprawling Italian-American family, Nicky was five years older and lived in another town. My image of this ball-playing, road-racing, gun-savvy, girl-crazy, practical-joking cousin had been built of brief glimpses at family gatherings and the memories of his buddies and family members. But suddenly, nearly three decades after he was killed, a chance discovery stripped away

3

everything I thought I knew about his death and drove me on a years-long quest filled with frustration and anger. Yet it would also hand me the gift of knowing my cousin as I never had when he was alive.

The actual circumstances of Nicky's death had been obscured for years. I went to Vietnam to get a sense of the place where it happened and had eleven days to follow his path, which is how long his fateful time in Vietnam lasted. If all went according to plan, I would find the very spot where an explosion cut him down.

As I traveled the straight and open road from Da Nang on that torrid day in May 1998, I saw that this was a trek into the past in more ways than one. We passed scenes that hadn't changed for centuries: peasants in conical straw hats ankle-deep in murky gray water, bent down to cultivate rice; skinny kids in shorts tending water buffaloes near thatch huts; women carrying baskets hanging from sticks across their shoulders; wooden carts waiting for their human haulers; cows plodding unheeded along paddy dikes.

I tried to picture the war superimposed on this pastoral place: Thirty years earlier, sweaty GIs in their steel helmets and olive-drab fatigues sloshed through the muck, dragging one jungle boot in front of the other, carrying rifles, machine guns and grenade launchers. From above came the *whop-whop-whop* of Huey helicopters—the kind Nicky learned to fly. They alighted and poured men out onto broiling terrain where the enemy was everywhere but on a front line, planting mines and booby traps and waiting in ambush. Viet Cong guerrillas could have been behind that hedgerow, or among those bamboo thickets, or in that patch of coconut palms across from the paddy fields, or around that hamlet.

Mien broke in on my thoughts, bringing me back to the present and calling my attention to a cluster of low buildings to the right, set back from the highway a few hundred yards.

"There, there!" he pointed. "It is where I went to school."

I saw a small, neatly kept campus, but if Mien hadn't told me it was still a school, I wouldn't have guessed. No children played outside, and the single-story concrete buildings looked like matchboxes.

Mien and I had met two days earlier at the Da Nang airport, and I felt lucky the travel agency had assigned him to me. He spoke English well and had the right connections to be a government-approved guide. Now on the road to Chu Lai, as we passed by his old school, I wondered about his boyhood during what his people called the "American War."

"Did the fighting affect you?"

"No, it did not happen here," he said. "This area we are passing, where I grew up, it was held by the soldiers of the Saigon government. They had control. It was safe."

That surprised me. I wouldn't have thought anywhere was safe.

As the car moved on and we drew closer to Chu Lai, I realized that my journey was nearing its goal. My stomach tightened.

~

Before I had left home, a friend who was a Vietnamese journalist had cautioned me about Chu Lai, saying it was a military base and the Vietnamese government considered it off-limits to foreigners.

"Visiting it is not quite easy," Le Quoc Vinh said.

He had taught me about the other side of the war when he stayed with my family for a month in the fall of 1993. At the time, he was a twenty-five-year-old editor with *Vietnam Investment Review* in his hometown of Hanoi, Vietnam's capital. He had come to my newspaper, *The Morning Call* in Allentown, Pennsylvania, under a Freedom Forum program that brought him to the United States for intensive training in journalism. One of my fellow editors announced Vinh would be coming to our newsroom and said the ideal housing arrangement would be with a family. My wife, Mary, and I volunteered to host him. We fit the bill, with two daughters, a couple of cats, a guinea pig, a parakeet and a spare bedroom in our nearly seventy-year-old, three-story Allentown twin. It would not be a first for us. We had opened our home to a young German journalist for three months after the Berlin Wall fell.

Vinh spoke English haltingly, dressed nattily, had jet-black hair and stood only five feet tall, his waist not much thicker than my thigh. Mary, who is just a few inches taller, found that hugging him was like cradling a delicate bird. Almost every night, he watched cartoons on our TV. They were easy to follow, he told us, and helped him with his English. We often heard him giggling in the living room.

One night he stayed up alone to watch the 1987 film *The Hanoi Hilton*, about North Vietnam's Hoa Lo prison, the former French colonial jail that held hundreds of U.S. prisoners, mostly downed fliers. The next day I asked him what he thought of the movie. "Oh, it was American propaganda."

Vinh's father worked for North Vietnam's news agency during the war and had gone to Hoa Lo to interview and photograph some of the prisoners. He concluded they were not so bad off. I wanted to tell Vinh that the brutality of "the Hilton" was well documented and that his father's visit had undoubtedly been well orchestrated. But I stayed silent, my duties as a host trumping my urge to be honest.

The day Vinh left Allentown, he said, "You must come to Vietnam for a visit." Mary and I laughed. We'd like that, we told him, but we were certain it would never happen.

Yet five years later, here I was on the road to Chu Lai, compelled by a mystery I could not ignore.

Days earlier I had entered Vietnam at Tan Son Nhut Airport in Ho Chi Minh City, formerly Saigon and still the country's commercial hub. About noon I checked into the Mercury Hotel. From my room on the seventh floor, I looked down on Tran Hung Dao Street, a broad boulevard fringed with trees and crowded with people who seemed not to notice the infernal heat. Riding motorbikes—*the* way to get around in one of the world's poorest countries—they flowed by like schools of fish, streaming past squat buildings emblazoned with the familiar names Toshiba, Opel, Suzuki. Many young women riders wore the traditional *ao dai*—pants and a long-sleeve tunic slit up the sides. Some looked like bandits, white kerchiefs over their faces.

The next morning, I took a taxi to the U.S. Consulate to explain why I had come to Vietnam and to provide a copy of my itinerary. The top official there raised a yellow flag about Chu Lai as Vinh had done, and noted that some Americans had been arrested when they tried to return to familiar places.

"If you get into trouble, we can't rush there and get you out of prison," Consul Ted Osius said flatly, as though by rote, words delivered to hundreds of other tourists who had come to commune with ghosts. "All we can do is encourage the Vietnamese to treat you properly."

That afternoon I met with Vinh, who now lived in Ho Chi Minh City. I had three days to spend with him before venturing north to Chu Lai. He was the editor of a slick magazine called *Beautiful Home*, a showcase for fine furnishings and architecture, and managing director of the company that published it. In his office, when I asked how he could sell his magazine in such a poor country, he said, "Well, I have 10,000 subscribers."

I got behind him on his motorbike, its motor blatted to life and in minutes we arrived at the War Remnants Museum, which I'd asked to see. Admission was 10,000 Dong, about seventy-five cents.

Outside, we walked among the relics of battle—a tank, a fighter jet, a helicopter. I was drawn to the chopper, a combat-scarred UH-1H Huey, one of the most familiar symbols of the war. Nicky had trained in Hueys when he attended the Army Aviation School at Fort Rucker, Alabama. I took in its sleek lines, plump cabin and tapered tail, and pictured Nicky strapped into the pilot's seat, flight helmet on. He had undergone rigorous schooling to earn his wings. But after he got to Vietnam, the place where he would have the chance to test his mettle, fate would not let him take the controls, not even once.

Inside the museum were pictures of American cruelty: napalm victims, the dead in the ruins of a bombed hospital, soldiers roughing up women in a village, dragging a man to his death behind an armored personnel carrier. A caption read: "After decapitating some guerrillas, a GI enjoyed being photographed with their heads in his hands." The embalmed remains of a deformed infant floated in a jar, labeled the victim of Agent Orange, a blend of herbicides that Americans sprayed to defoliate the jungle hiding places of their foes.

I didn't ask what Vinh thought about these displays. We didn't talk at all. He seemed indifferent, maybe because he'd been there before or maybe because he didn't want me to feel uncomfortable. I was, a little, but I wasn't surprised at what I saw. North Vietnam won the war, so

My friend Le Quoc Vinh, a Vietnamese journalist, poses in front of a UH-1H Huey helicopter at the War Remnants Museum in Ho Chi Minh City, the former Saigon, in May 1998. Vinh was then a magazine editor. As he and I toured the museum together, I snapped the photograph.

of course the leaders of a unified Vietnam wanted to shame the losers. But over the years, the government had softened its attitude toward the United States. This hadn't always been the War Remnants Museum; it used to be the American War Crimes Museum.

Afterward, Vinh and I went for iced coffee in a café, where I was startled to see a gecko racing up the wall just behind him. "This is Vietnam!" Vinh said with a laugh when I pointed at it. I spied several more that night on the walls of his family's apartment, where we had dinner. They weren't such awful intruders, I learned. Two Irish cousins I'd met in a Ho Chi Minh City restaurant told me that the lizards in their budget hotel didn't bother them nearly as much as the large rats.

The next day, a guide took me to see the abandoned U.S. Embassy, where in the spring of 1975 choppers hastily evacuated the last Americans as communist troops closed in. It was a bleak reminder of failed policy, a monument to squandered lives and treasure. It would be torn down soon after I left the city.

We moved on to Independence Palace, now called Reunification Palace, which had been the seat of South Vietnam's government. From inside I looked out at the front gate and realized I'd seen this place before. More than two decades earlier, TV news footage broadcast over and over showed a North Vietnamese tank crashing through the gate, dramatizing the fall of South Vietnam and America's final humiliation.

When the communists swept into Saigon on April 30, 1975, I was a junior in college and co-editor of the school newspaper. We bannered the story across the top of Page 1. As I looked over the page before it went to the printer that night, I thought of my cousin. His life had been sacrificed, and for what? It reminded me that I had the chance Nicky didn't get—the chance to go on. It was one of a handful of times that thoughts of my lost cousin surfaced as I grew into adulthood.

At the end of my three days in Ho Chi Minh City, I said goodbye to Vinh, who today is the master of a media empire in Hanoi that includes five magazines, a cable TV channel and a marketing and communications company. I fretted a little on my flight north to Da Nang, the beginning of the crucial leg of my journey, about whether I'd get into Chu Lai. But my guide, Dinh Mien, quickly put my concerns to rest after greeting me at the airport. He knew the region and had been to Chu Lai before, most recently when he brought a former U.S. Army nurse onto the base and helped her find a Vietnamese woman she had tended in the late 1960s. Perhaps most important, he had cleared my visit with a commander in Vietnam's army.

The next morning, as I waited in the lobby of Da Nang's Faifo Hotel for Mien and our driver, a fiftyish French tourist asked why I had come to Vietnam. I told her about Nicky and showed her a picture of him. She left the lobby, came back with a daffodil and placed it in my hand. "You have touched me with this story," she said. "I want you to have this in your cousin's memory."

That was more than an hour earlier. Now we were approaching Chu Lai. To our left, lush land turned sparse and sandy. We were nearing the South China Sea.

Mien's plan when we got to the former U.S. base was to start by looking for the firing range at Landing Zone Bayonet. That was where Nicky and other soldiers newly arrived at Chu Lai were trucked for part of their weeklong orientation. A one-story, plywood building alongside the range had served as a classroom.

"David, Chu Lai," Mien said.

If he hadn't alerted me, I might not have known. Just sun-bleached hard dirt, scrub brush, scattered trees and gently sloping dunes of white sand that looked almost like snow lay where the U.S. military base had been years before. Drooping, crisscrossing power lines and the poles that held them aloft were the only evidence of humans on this parched piece of coastal plain.

In 1969, Chu Lai's three dozen square miles teemed with 25,000 fighting men and support personnel, a key base for Marine Corps fliers and the Army hub of operations for the Americal Division, to which Nicky belonged. The Air Force had personnel here as well and the Navy ran a shallow-water port, the second-busiest port in the region after Da Nang's. Now nothing—no rusting Quonset huts or aircraft hangars, no junked two-and-a-half-ton trucks or cannibalized choppers. After the Americans departed, Mien said, peasants stripped down the buildings and carted away the parts. They peddled what they could or used the pickings for themselves.

We drew near the main gate. It stood by itself in the open, a neglected monument, no fence, no wall and no one in the booth. Above it were the words CAN CU, signifying a base, and in larger letters, CHU LAI.

I reached for my camera.

"No pictures! No pictures!" Mien cried when he saw my movement out of the corner of his eye. Not even Vietnamese were permitted to photograph the gate, he told me gravely. Taking photos of military installations was prohibited and could lead to harsh penalties.

"You would be in trouble, and so would I."

"Who's going to know?" I asked. "There's no one around."

"They'll know, they'll know!" he said, waving his arms in agitation.

I put down my camera. No sense risking a problem over something so minor.

We went a little farther, pulled off Highway 1 to the right and stopped beside a dilapidated wooden shack next to a blood-red sign for a quarry company. Mien got out of the car and went inside with my most detailed Army map, hoping to find someone who knew the location of the wartime firing range. The U.S. military had kept the Vietnamese from living in the area, Mien had told me. But after the war, a hamlet popped up across the highway from the base.

Mien came back, unsuccessful. As we pushed on, I lost track of time as our car clambered over dirt paths and Mien questioned anyone we met—two boys on a bicycle, a shirtless old man in a straw hat, a bespectacled worker out in a clearing where an open-bed truck was parked, youngsters and other villagers mingling in front of a one-story school.

We drove over a barely passable dirt trail that led to a wide-open area dominated by a stone-crushing machine, the quarry announced by the sign. Poking out of the machine were conveyor belts that slanted up and down. Beneath them lay gray mounds of the finished product— small, squared stones. Workers rested alongside the site, lounging in the shade to avoid the punishing mid-morning sun.

Mien spoke with some of the men. Yes, we were on what had been LZ Bayonet, home of the Americal Division's 198th Light Infantry Brigade, but we weren't near the firing range. It was somewhere in the direction from which we'd just come, though no one knew exactly where. We got on Highway 1 and headed back. Manh pulled over to a restaurant in the pleasant shade of mango trees. Out front, a young man picked through a girl's shoulder-length hair for lice. A scrawny black dog rested on its haunches at her feet. Young men ate noodles in the steaming open air beside a table piled high with watermelons. Mien came out of the restaurant with a mustachioed man in a cap, a dark blue polo shirt and faded, torn, baggy blue jeans. He got into the front seat with Mien.

"He knows where the firing range was," Mien said. "He will show us."

Our new guide, who once served with the Army of the Republic of Vietnam, remembered the range because he had ridden past it during the war. He directed us over a bumpy dirt trail. We didn't go far. In less than fifteen minutes, we stopped. A wall of savage heat hit me when I got out of the car—heavy, smothering air that closed in thickly around me. For the hundredth time I wondered how our troops endured this vicious climate, lugging weapons, ammo and gear, trying to do their job and stay alive.

But it wasn't the sultry steaminess that caused me to suck in my breath abruptly and hold it. It's what I saw.

To my left was a long, level expanse of baked earth that approached a rocky, green ridge of high hills a few hundred yards away. I had seen this before in an Americal Division historical video showing men shooting toward the hills.

"This is the firing range," Mien said.

I wandered toward the hills, imagining bullets ricocheting and sending off fierce sparks. Mien went in the opposite direction, toward a farmhouse and other closely packed buildings. Sweat slithered down my chest and legs, soaking my cotton clothes.

A compass showed I was north of the hills, which meant that the building where classes were held would have stood behind me and to the right. But there was no sign of a building.

Damn, I had to know exactly where it had been.

"David, David!" Mien called suddenly from the farmhouse. I turned. "The classroom!" he shouted and pointed toward a patch of eucalyptus trees and brush. An old farmer carrying a straw hat was with him.

Mien said the farmer remembered the classroom because he had been stationed nearby while in the military. After South Vietnam's defeat, he had moved here to grow rice and sweet potatoes. The classroom, the farmer recalled, was a big building.

I knew its size—twenty-five by fifty feet. It was made of plywood and had a corrugated tin roof. But there was nothing left of it. The farmer, with Mien interpreting, said peasants had dismantled the building and carted away the materials, just as they had done at the base across the highway.

A deteriorating asphalt road ran between the edge of the firing range and the classroom site. It was the road the trucks from Chu Lai used to bring Nicky and other soldiers to the orientation building.

By now, some rail-thin men and boys had gathered around us, curious. They gaped at me when I spoke into my microcassette recorder, describing my surroundings. They might have thought I was talking to myself.

I knelt down on one knee among eucalyptus trees, scooped a handful of sand and watched it run through my fingers. I pressed my hand flat on the ground where the sand fell.

Twenty-nine years earlier, Nicky's blood smeared this earth.

I looked up, imagining the approach of Army trucks. They stopped in front of me and a few dozen soldiers hopped out, Nicky among them. They strode through the screen door into the classroom. I heard a man's voice but couldn't make out what he said. An earsplitting explosion convulsed the building. Men screamed, shouted, ran in terror. Someone pulled Nicky out. He lay crumpled on the ground, going into shock.

My heart thumped. My eyes, burning from the threat of tears, scanned the ground. I wanted to remember how the sand looked—brownish, not the bright white of the Chu Lai dunes—and how it felt sifting through my fingers—soft, grainy, filled with bits of brush, warm because the trees shaded it from the brutal sun.

I had come halfway around the world to stand on this spot.

Vietnamese children join me near where the orientation building once stood on the former site of LZ Bayonet, outside the onetime U.S. base at Chu Lai, Vietnam, in May 1998. My guide, Dinh Mien, helped me find the spot, which is approximately where an explosion cut down my cousin Nicky.

Chapter 1

That Dirty War,
July 28, 1969

I was fifteen when Nicky came home in a silver metal casket. The day he was buried, rain fell in the morning. It had stopped by noon, when the procession of cars from Malvern filed into the hundred-acre cemetery of slight hills five miles away. My parents, my two brothers, my sister and I joined the crowd by the grave, which lay alongside my grandmother's and in the shadow of three clustered magnolia trees.

The grass was wet, the moist air thick with grief. From the stillness and silence of those around me, I immediately understood that I was standing in the midst of an uncommon sadness, something I had never felt emanating from so many people together at one time. Its gravity unsettled me. I became acutely self-aware, concerned that if I didn't remain perfectly still and quiet like everyone else, I might disturb the tableau. As a member of Nicky's extended family, I was part of this deeply respectful gathering, yet detached. His loss didn't hit me nearly as much as it had others, because he and I had never had more than fleeting contact. I only knew that he was my cousin and that he was killed in Vietnam shortly after getting there.

An Army officer in dress uniform stood near me at attention, eyes fixed ahead. A tear glistened on his cheek. It struck me how Nicky's death could touch a stranger's heart, how it had moved this man who otherwise had the look of hard discipline.

The soldier was Second Lieutenant Victor J. Elmer from Marksville, Louisiana, an engineer stationed at Valley Forge General Hospital, and he had been with Nicky for several days. He had escorted the body to Pennsylvania in a hearse from Dover Air Force Base in Delaware, that well-known portal through which the military dead return in flag-draped coffins. Nicky had arrived at Dover almost a week before the funeral—and not alone. Another body on the plane was that of his friend and fellow helicopter pilot Billy Vachon (pronounced VASH-ahn), who died with Nicky. They were two of 11,614 American deaths in Vietnam in 1969—a terrible toll trumped only by that of 1968.

Billy, the father of a toddler, was going home to Portland, Maine.

My parents had not brought my younger brother and sister or me to Nicky's viewing the evening before. Mom and Dad, my older brother and the others who did go to Mauger's Funeral Home that Sunday saw him laid out in his Army uniform in a standard military casket. Under the lid was a metal liner that went up to Nicky's waist, preventing anyone from seeing that his left leg below the knee was missing. From the waist up he was enclosed by glass, which distressed some of the mourners who felt deprived of a final moment of intimacy. No hand could touch him to say goodbye.

Nicky's eighteen-year-old fiancée, Terri Pezick, approached the casket in a blue-and-white dress he had especially liked, steadied on either side by two members of her family. She carried her high school ring, which Nicky had worn on his pinky and had left with her when he departed for Vietnam. It was attached to a white, heart-shaped ribbon. She placed it between the glass and the liner so that when the lid came down, the symbol of her love would be enshrined with him forever.

Monday morning, we attended a Requiem Mass at St. Patrick's Church, where Nicky had gone with Terri right before his deployment. The pews were packed with mourners, many of whom I didn't recognize. We went on to Philadelphia Memorial Park in Frazer for the graveside service, led by a newly ordained Catholic priest, the Rev. John J. Kilgallon, twenty-six years old. He said the prayer of committal, the Rite of Christian Burial.

But on that day, July 28, 1969, we did not witness the ceremony owed an American killed in the war. There was no meticulous folding of the flag for presentation to Nicky's mom, no crack of a rifle salute, no solemn finality in the bugle call of taps. The Army had failed to send a military burial detail. An apology eventually would come from the garrison commander at Fort Indiantown Gap near Harrisburg for what he called a "communications discrepancy." The bureaucratic blunder presaged what I would discover decades later about Nicky's death.

After we prayed that July day and filed past the casket, it was lowered into the grave and Nicky was gone except in memory. The earth that covered him would soon be topped with a Veterans Administration plate identifying him as a warrant officer in the Americal Division, Vietnam, and bracketing his life: November 26, 1948, to July 15, 1969.

We gathered afterward at the Malvern home of my dad's sister Patty and her husband, Jimmy. Their house was the heart of the family, inherited from Grandmom and Grandpop, who had raised the youngest of their dozen children there decades earlier. Built about 1887, it was a wood-frame structure two stories high, later sided with stucco and enlarged with an addition in the rear. It stands close to other old homes on King Street, the main road through town, running roughly parallel to the two sets of tracks of the old Pennsylvania Railroad. From the street, concrete steps lead to a porch and the first floor. A room in front at sidewalk level had once been a shop where shoes were sold and repaired.

Malvern was where Nicky had grown up. Twenty-seven miles west of Philadelphia on the suburban Main Line, it was a picture of small-town America. Its people worked in mills, markets, and stores or ran barbershops, garages and eateries. Parades filled King Street every Halloween and on the Sunday after Memorial Day. Summertime meant fire company fairs, baseball and community picnics. A monument to fifty-three Continental Army soldiers killed in the Revolution's "Paoli Massacre," a British thrashing of the rebels that happened in 1777 within the borough's one-and-a-third square miles, reminded visitors of the price of freedom and the town's deep roots in the American story.

My grandparents moved to the house as renters in 1938, while the Great Depression wracked the country. They had been living in a stone farmhouse less than two miles away, close to where Grandpop and his older sons wielded picks in a quarry for the Chester Valley Lime Company. The new place had amenities the family wasn't accustomed to—electricity and running water. Five years later, Grandmom and Grandpop staked their claim to the American dream, buying the house from a bank for $2,500.

All my life I had been coming here for holidays and happy times. As with most Italian-American families, dinner itself was a rousing celebration. Grandmom, a busy cook and baker, would serve big pans of chicken and potatoes, pot pie with raw onions, pizza almost two inches thick.

This day was different. Among adults, grief melded with anger. Uncle Jimmy railed against "the damn government and its dirty war in

Vietnam." His rage did not spring from the same source as the street protests by the left-wing college kids of the day. My family was the epitome of patriotic, working-class America and had quietly gone about its business through the turbulent 1960s.

A generation earlier, World War II had galvanized the Vendittas' patriotism. Grandpop had come to America from Italy in 1903 as a twenty-year-old of little means and had been grateful for a better life in a land of promise. Four of his sons—including my father and Nicky's dad, Louie—helped fight Germany and Japan. Uncle Jimmy served in the Army during the Korean War.

But now the grinding tragedy of Vietnam had struck my family in the form of Nicky's loss and turned these ordinary Americans against their country's involvement in a small Southeast Asian nation's conflict. Several adults vowed to send us kids to Canada if necessary to keep us from fighting in a faraway jungle for a dubious cause. No one else was going to end up like Nicky. No one else was even going to come close. In my fifteen-year-old mind, I saw myself furtively taking a bus to a land of snow and ice, dark forests, large animals and mounted police in bright red jackets and shiny black boots. It would be a vast unknown, an adventure.

The anger and grief of that day stuck with me, triggered months later by something as mundane as an assignment in a typing class at school. Mr. Angle had told us to write a formal letter to anyone we chose. My letter was to President Richard M. Nixon. It was brief, blunt and artless. I told him I didn't like the way he was running the country, that my cousin was killed last summer in Vietnam and that Nixon was responsible.

Mr. Angle stopped by my desk a few days after I'd handed in the assignment. He leaned over close to my ear and said: "That was no way to write to the president." His face was stern, his jaw set tight. He didn't wait for me to answer, just stalked up the aisle. I looked down at my desktop, my face burning in shame.

Days later, humiliation turned to annoyance. I was in tenth grade, I was a citizen of a free country and I could say whatever I wanted to the president, however I wanted.

My lack of respect for the man holding the highest office in the land didn't hurt me at the end of the semester. Mr. Angle gave me a good grade. I also received a large manila envelope from the White House, no sympathy note included. The envelope contained copies of two speeches Nixon had made about Vietnam. On November 3, 1969, he asked that people support him while he worked toward "a just and lasting peace." The next month, referring to the massive antiwar demonstrations across America, he said North Vietnam should not expect "division in the United States would eventually bring them the victory they cannot win over our fighting men in Vietnam."

Neither speech impressed me. Nor did anything Nixon or anyone else in authority said or would say about the Vietnam War. I retreated into adolescent indifference and went through high school shutting out the war and the conflict it engendered.

Those days were carefree and fun. I spent them performing in plays, musicals and the school choir, going to football games with friends and dating a spirited, popular girl. When a senior, I edited the weekly student newspaper and don't recall a word about the war getting into it. In a social studies class in the spring of that year, our teacher talked about Vietnam and said some of us might go there. I briefly wondered if Mr. Nippes could be right, but then brushed it off, thinking it would never happen to me.

I didn't know enough to understand that the war was winding down and he probably was wrong. Only 70,000 Americans remained in Vietnam that spring of 1972, down from a peak of more than half a million. Then, two months after I graduated, the last U.S. ground combat troops pulled out. By then I was busy with my summer job cleaning toilets and mopping floors at a local country club, excited about going to college in the fall. I missed the news of the withdrawal.

Nicky too faded deep into the background. He had always been a remote figure, the older son of my Uncle Louie and his first wife.

In the early 1960s, Nicky lived with his mother on East King Street in Malvern, a town of some 2,200 residents. Occasionally my dad would drive us there in our long black '59 Ford station wagon, which ran like an armored vehicle. Coming east from our home outside Downingtown, Dad would steer off Route 30 onto the Old Lincoln Highway, curve up the ridge and turn right onto Bridge Street. We'd cross the hump-like bridge over the railroad tracks and plop down to the "T" intersection with East King. There he'd turn left and we'd arrive at Grandmom and Grandpop's house.

A right turn off Bridge Street would have taken us, in less than a block, to Nicky's house, where the backyard reached the sunken rail bed. We had no reason to visit Aunt Sally, who was Nicky's mom and divorced from Louie, so we never went there. Trips to Malvern centered on holidays or occasions like my grandparents' birthdays. Sally didn't attend.

Nicky did turn up at family get-togethers where I was also in the crowd. Two images of him linger on the edge of my memory like frames frozen on a video screen. In one, I am about twelve, it is summer and we are at a family picnic outside Malvern. Nicky and another cousin are standing beside each other. All I see is their faces, but only Nicky's is in focus. He's looking down at me, grinning and saying "hi." The warmth in his voice and in his hazel eyes tells me his friendliness is genuine. But I'm so surprised he's spoken to me—a mere little kid—that I don't say anything other than a weak "hi" in return. I gape at him, and he keeps smiling, his eyes never leaving mine.

The second is from another summer picnic outside Malvern a few years later. I am fourteen. It's at the home of Nicky's dad, Louie, and his second wife, Bert. They have invited friends and relatives over to wish Nicky good luck as he leaves for boot camp down south. We are all in the front yard. Nicky's girlfriend is with him. I'm a dozen feet away, talking to someone else, when I happen to turn and my mind's eye snaps the picture. In this frame, I see only Nicky and the girl, holding each other in the sun and laughing.

How can I fetch from distant memory two pictures as common as a simple greeting and a young couple embracing? It was as if I knew to take notice, recognizing that someday I'd need the images to connect me to Nicky again.

Miles from Malvern and twenty-five years after Nicky's violent death, that day came.

Chapter 2

Discovery, November 1994

On the Sunday after Veterans Day 1994, I was working at *The Morning Call* in Allentown. The newsroom was quiet except for the occasional crackle of the police scanner. I paged through the previous day's paper looking for story ideas, nuggets we could use in the days ahead.

A headline caught my attention: "Local Vietnam Vets Search Locator Service to Find Friends." The story was about veterans using a new computer database on all Americans killed or missing in action in the Vietnam War. It had birth dates, hometowns, years of service, the date, type and cause of casualties and where they occurred. It gave the location of names on the Wall in Washington, D.C.

I wondered what they had on Nicky. I cut out the story.

In those days, I had no Internet connection, so I couldn't use my home computer to access the database. Instead I phoned the Friends of the Vietnam Veterans Memorial, the nonprofit group that owned the data, after the story had sat on my desk at home for a couple of weeks. I gave Nicky's name to a woman who answered and asked her to mail me his information. I didn't expect anything dramatic. After all, I'd known for years from family members what had happened to him: Just days after he arrived in Vietnam, he was with other guys waiting for a ride somewhere. An enemy rocket exploded among them. Nicky lost a leg, lived several days and died.

That was it. A rocket got him.

He had been my only relative who went to Vietnam and the only soldier I knew who didn't come home alive. For more than two decades, whenever the war came up in conversation, I had told people about him and how he was killed, because it was something I could offer.

A few days after my call, a one-page printout arrived in the mail. It identified Nicholas Louis Venditti as a warrant officer in the Army Reserve. He was born November 26, 1948. He was a single, Caucasian Roman Catholic from Malvern, Pennsylvania. His tour of duty started July 3, 1969, and he died a dozen days later, on July 15. His name was on Wall Panel 20W, Line 3, of the Vietnam Veterans Memorial. All as expected.

Then, Casualty Type: "Non-hostile, died illness, injury."

Cause: "Accident."

What? I put the paper down and picked it up again, thinking I'd read it wrong. How could Nicky have died in an accident? What kind of accident?

It was a weekend, so I had to wait until the following Monday, November 28, to call the Friends group again. "Exactly what does this mean?" I asked the woman who answered. She said it meant Nicky didn't die at the hands of the enemy.

Who did kill him? How could I find out more?

She gave me the number for the Army's Casualty and Memorial Affairs Operations Center in Alexandria, Virginia. I called and spoke with mortuary program specialist Doug Howard. He promised to look up Nicky's file and get back to me. That afternoon, he did. He called me at work and read from the file: Nicky had been fatally injured "while in classroom at base camp when the instructor of class accidentally detonated live grenade."

"What? What else does it say?"

Nicky was severely injured and taken to a base hospital, where his left leg was amputated below the knee. He was transferred to another hospital on the same base, dying there five days after the accident.

I pressed Howard to tell me if there was anything more about the explosion. No. But he offered to send me a copy of the casualty file. "It could take four to six weeks," he said.

Incredulous about my discovery, I immediately told some of my co-workers. They shook their heads in disbelief. How could such an accident happen, we wondered, and how did the circumstances of his death get lost in the telling?

My dad, Nicky's uncle, would not be able to help me; Alzheimer's disease had eaten away his memory. Mom was caring for him at home as best she could. I called and asked her how Nicky died. He and some other soldiers were waiting for a transport when a rocket hit, she said.

"That's not what happened, Mom," I said.

"What do you mean?"

When I told her what I'd learned that day, there was a long silence on the other end of the line.

"Where did you hear Nicky was killed by an enemy rocket?" I asked.

Mom couldn't remember.

She and Dad had a scrapbook with a clipping from the local newspaper. According to the story, Nicky died "as the result of wounds suffered in action about a week after he arrived in the war zone." He'd been in a war zone, but he hadn't been wounded in action. Somehow, the *Daily Local News* of West Chester had gotten it wrong.

I needed to know more about Nicky's fate. I tried and failed to picture how the accident happened, lying awake at night for weeks imagining the moment of the explosion without the details to fully grasp it.

More than two months passed before I got the casualty file, about thirty pages of reports, telegrams and letters. Mary and I sat together at the kitchen table and leafed through it in distress. It gave the extent of Nicky's injuries: fracture of the left femur, multiple fragment wounds to the right arm and both legs, surgical amputation of the left leg below the knee. One page had an anatomical chart showing two drawings of a man, front and back, with lines pointing to the location of Nicky's wounds and a corresponding description: four points on his right arm and hand, *mutilated*; eight points on his right leg, *mutilated*; three points on his left leg, *mutilated*. Below the left knee, the leg was blacked out and labeled *missing.*

The file states Nicky died from fat microemboli, cerebral and pulmonary. Microemboli are microscopic droplets of oil or fat that can appear after injury to a long bone. They had probably leaked from the marrow of Nicky's shattered femur, traveled through his bloodstream and blocked blood vessels in his lungs and brain.

I ached at knowing how Nicky had suffered in his last days.

Our bewilderment only grew months later when I received another document, Nicky's military personnel file, which clearly shows that his parents knew the circumstances of his death. A telegram sent to Uncle Louie on July 12, 1969, two days after Nicky was hit, states he had been hurt "while in a training session when a grenade accidentally detonated."

All that information emerged, down to how much money Nicky had on him—$254.80—yet neither file shed any light on what I was burning to know: how and why the explosion happened. The files concerned only Nicky. They had nothing that revealed the larger picture of an Army instructor's fatal mistake. He had somehow set off a grenade in a classroom. What had he been doing? Other than Nicky, how many men had been killed or wounded? Had the Army conducted an investigation? Did the instructor survive the blast? Was he held accountable and punished? Was anyone else implicated, such as a superior officer?

I could not have understood then that my hunt for answers would take decades.

Fortunately, a career in journalism born of a love for writing and storytelling had given me the skills I would need to carry out the search and the resolve to press on. I'd been committed to newspapers since high school, when I realized it was what I wanted to do—demanding, exciting, meaningful work that served the community, made a difference. It had a romantic appeal I might have recognized even as a two-year-old—a family photo shows me in pajamas cross-legged on the floor, earnestly tapping on a toy typewriter I'd gotten for Christmas. When I got into trouble for some misdeed at St. Joseph School in Downingtown, a nun ordered me to write 500 words on "Why I Must Be Obedient." She liked my little essay so much she read it aloud to our class, as I sank down in my seat. At the public junior high a few years later, I won first place in a tall-tale–writing contest in Miss Orazi's seventh-grade English class. High school and college newspapers beckoned. After graduating from Indiana University of Pennsylvania in the bicentennial year, I got my first full-time reporting job at a small daily in the Alleghenies. Later I came east to a paper in a Philadelphia suburb, then headed north to Allentown.

That was the end of my moving around. Mary and I met and fell in love in *The Morning Call* newsroom, where we both worked at night, and we were married the following year, 1986. In time, she took jobs elsewhere in public relations and graphic design, while I was happy to stay at the Lehigh Valley's largest newspaper.

By 1994, I had been a professional writer and editor for eighteen years. Looking back, it seems as though all the work I'd done had prepared me for the biggest, most challenging story I would ever tackle—reconstructing Nicky's short life. I would discover what motivated him to join the Army in 1968, the deadliest year for Americans in Vietnam: The military would teach him how to fly, and after a one-year adventure in the war and his service was done, he could make a good living as a commercial pilot. He would marry his girl and have a bunch of kids. He had it all figured out.

It would have been terrible if he had been killed when his chopper blew up in the sky, or was riddled with gunfire and crashed into jungle canopy, or if, while it was hovering low over a landing zone, he had been shot. It would have been tragic if he had died when a rocket landed on the base as he slept, or if his aircraft had plunged into the sea after a mechanical failure. Still, any one of those scenarios would have been understandable.

But I couldn't comprehend why his life had been wasted in a classroom accident.

As I pored over the paperwork at our kitchen table, I realized I had never heard Nicky's parents talk about what happened to their son. Mom said for them it wasn't how Nicky died, only that he was gone.

I wondered if I could muster the sensitivity and tact to speak with them. Years before, I'd made a stupid, offhand remark during a family party within earshot of Nicky's dad. Finishing a beer, I crumpled the can in my hand and said, "Well, there's one dead soldier," quickly realizing that Uncle Louie was close by and probably heard me. I searched his face, but he avoided looking at me. Shame flooded over me at the memory.

Despite that, I had to try.

I wrote to Uncle Louie and Aunt Bert, saying I'd like to meet with them and talk about Nicky so I could write about him. "It's important to me to remember him," I said. "He was my cousin, he went to a war that happened in my time, and he was so quickly lost." My journey of discovery would result in a heartfelt memorial to their son. I ended by saying I'd call them in ten days.

That sealed my commitment to do something more than just think about Nicky and collect documents. I had to know more about him and what happened to him in Vietnam, at least so our family would have the details.

The afternoon of November 26, 1995, the date that would have marked Nicky's forty-seventh birthday, I stood at the wooden desk in my home office—a third-floor cubbyhole with windows facing a cemetery—and started to phone Uncle Louie and Aunt Bert but kept picking up the receiver and setting it down. Once, twice, three times. Who would answer? What should I say? But it was too late to back down. They were expecting my call.

Aunt Bert answered the phone.

"Did you get my letter?" I asked.

"Yes."

"Will you and Uncle Louie talk to me about Nicky?"

A pause, then "Yes."

"Could I come visit you in two weeks?"

"That would be fine."

Chapter 3

"A good kid,"
December 8, 1995

Uncle Louie carried a cardboard box into the kitchen of his modest ranch home outside Malvern and tipped it, spilling the remains of his son's life onto the table.

Family photos, Nicky's orders to Vietnam, silver wings and other pins, a pilot's checklist, ID cards, a safe conduct pass, a silver bracelet with "Terri" etched on the smooth face, all sent home after Nicky's death. They were mixed with items my aunt and uncle had saved, including the Western Union telegram informing them of Nicky's grievous injuries, still in its faint yellow envelope with the slogan "On any occasion, it's wise to wire" on the back.

"I've only gotten this out once or twice before," Louie said, and stepped back to glance at Aunt Bert, who had made a pot of coffee. She filled two cups and sat with me at the table against a wall decorated with a wood block saying "Gone to bingo" and a sign, "Old friends are the best antiques." It was December 8, 1995, and though Christmas was only two weeks away, holiday cheer was absent; they had no decorations up. At my feet, their black cocker spaniel, Sparky, pawed at my satchel.

Louie stood by the back door tentatively, avoiding my gaze. He was short, only five feet six inches tall. As a younger man, he'd had an erect posture, thick forearms and a scrappy build that suggested coiled energy, which is why he had always reminded me of a spark plug. He still had the square stance, but the brown had ebbed from his curly hair and he was thinner now, in his early seventies. He looked tired, his face wan and the glimmer gone from brown eyes that used to flash with fun and mischief.

Bert, the chattier of the two, had an easy and oftentimes sly good humor that broke through in a husky laugh. Unlike Louie, who rarely ventured beyond the familiar haunts of his hometown, she liked to go places. She and my mom had rollicking-good times together at the casinos, playing the slots. As with Louie, this was Bert's second marriage. Her first husband was a painting contractor with whom she had lived in Malvern and had a son four months younger than Nicky. Seventy now, her once blond hair in silver curls, she stared—silent and somber—at the remnants of her stepson's abbreviated life scattered in front of us.

I picked up an aged wallet, its brown cover torn and gouged. "He had that with him when it happened, when the grenade went off," Aunt Bert said, nodding toward the wallet as I rubbed my thumb over the goatskin. "It's got blood on it." But I couldn't see blood. Maybe age had melded red into brown.

Inside the wallet was an undated clipping from the *Daily Local News* noting Nicky's engagement to Terri Pezick, a student at Great Valley High School. A smaller-than-wallet-size portrait of Terri showed her short, light brown hair swept across her forehead and her thin lips curled in a half-smile. On the back, she had written in blue ballpoint, "I love you very much for always. Love you always."

Cut to fit into Nicky's wallet were pictures of Terri caught in everyday life: kneeling beside a poodle Nicky had given her, posing with two little Dachshunds in her lap, ironing, sitting at the kitchen table as Bert dipped a spoon into a bowl of soup. Another showed his dad mugging in a hardhat.

"He was a good ballplayer," Uncle Louie said when I asked him what he remembered most about his lost son. "He started in Little League, then Babe Ruth and American Legion. He played third base and pitched. He pitched some of a championship game in Phoenixville that his team lost, and that discouraged him. He didn't have the drive to play after that."

Abruptly my uncle retreated into the dining room out of sight and said loud enough for me to hear, "He was a good kid." In a moment he was back in the kitchen, but he looked away from me.

When Nicky turned eighteen, he moved to this rancher on King Road from his mom's Victorian house just a few minutes away. He had wanted to come here sooner to be with his stepbrother, Joe Gray, Bert's son. The two were close and shared interests in baseball, guns and cars. But Sally, Nicky's mom, insisted he remain with her. Under the terms of the divorce, she had full custody until he became an adult, and she valued the child support Louie paid monthly. "When I'm eighteen, I'm moving down with Joe," Nicky told a friend. It happened in November 1966, uniting father and son under the same roof for the first time in eight years.

Moving in with his dad strengthened Nicky's relationship with his stepmother, who had known him since he was in grammar school and treated him with humor and affection. Nicky felt comfortable with Bert, and she with him. That fondness grew with trust, which led Nicky to confide in Bert his trysts with girls and to seek her advice on money and morality.

One day, Nicky announced that he wanted to be an Army helicopter pilot.

"I didn't believe him," Uncle Louie told me, tears welling. He slipped out of the kitchen again.

"Why not?"

"Just didn't," he called to me from the dining room. "I thought he was kidding me."

But Louie didn't object. Aunt Bert did. She had seen TV images of the peril that helicopter pilots faced over enemy-held hills and jungles. Low and slow, the choppers made good targets.

Why did Nicky want to fly?

"I don't know," Uncle Louie said, returning to the kitchen. But then he brightened. "Maybe it was because I used to tell him World War II stories about the Army Air Corps."

Anyone willing to listen knew that Uncle Louie had gained a deep respect for pilots while a ground crewman with the 8th Air Force in England, driving fire trucks and other vehicles for a fighter squadron. Now that he had brought up the war, he wanted to talk about the men of his family.

"Did you know that five of us were in the service?" he asked, showing me a paper titled "Honor Roll of Malvern Borough." It was a list of the 204 residents who joined the armed forces or the maritime service during World War II. They all were there, the patriarchs of my family: brothers Sam, Frank, Louie and my dad, Carmine. Their older brother Jimmy was killed by a drunken driver in 1934, but his son—also named Jimmy—was on the list. "We called him Junior," Uncle Louie said.

Sam, Frank and Junior were in the Army, and my dad was in the Coast Guard. All made it through unscathed except Sam, who suffered a brain injury while serving with the Coast Artillery in the South

Pacific. It happened while he was blasting coral on the island of Bora Bora, subjecting him to a continuous concussion of explosives over many hours. Doctors determined he'd had epilepsy before the war and the blasts aggravated his condition. Seizures dogged him for eight years until one killed him.

But he lived long enough to dote on his nephew. A photo shows Uncle Sam happily holding one-year-old Nicky on a tricycle. Stalking them both was the specter of life cut short by the accidents of war.

On the way to my aunt and uncle's house that December day, I stopped by Philadelphia Memorial Park to visit Nicky's grave. It was a familiar place. My father had worked at the cemetery office as an accountant in the 1970s, and years before that, his father had tended the grounds. An older man in the office who remembered my dad pointed out where Nicky lies. It was close by, shaded by the three magnolias.

I saw the bronze Veterans Administration plate set flush with the ground. Oxidation had crusted it to a greenish blue. Beside the plate was a small American flag Uncle Louie had placed there. He kept plenty of extra flags and would replace the one that flew in tribute to his son whenever it looked worn or faded.

Nicky is buried next to our grandparents, Nicola and Mary Cugino Venditta. Grandmom, born in Philadelphia and married on her fifteenth birthday, died during Nicky's senior year in high school. Grandpop, who had emigrated from central Italy's Campobasso Province and spoke little English, outlived Nicky by two years.

Nicky Venditti as a one-year-old on a tricycle with his uncle, Sam Venditta, one of my dad's older brothers. The photograph was taken in their hometown of Malvern, Pennsylvania. In May 1950, Sam died of a World War II noncombat injury. The difference in the spelling of their surname was a quirk of the family (courtesy the collection of Sally and John Pusey).

"Hi Nicky," I said. "I know what happened to you. Now I want to get to know you. I'm going to see your folks."

At the house later that day, Louie stood under a rusty horseshoe propped on the back-door ledge in the kitchen. He had always been an affable, playful rogue. Drawn to beer and his buddies, he whiled away many nights drinking, throwing darts, shooting pool and swapping stories at the Paoli Veterans of Foreign Wars' post down Grubb Road—"the Vets," as the locals called it—or at the Paoli American Legion post. He could entertain folks for hours with a cascade of yarns and jokes.

Aunt Bert called him an instigator, a trait he passed on to Nicky. "Louie would come home from the Vets after only being there about an hour," she said. "I'd ask him what's up, and he'd say people there were arguing, and that's because he'd started the argument, and then he'd just leave there laughing."

Nicola and Mary's seventh child, Louie was born in 1923 in central Pennsylvania's Mifflin County, amid the Ridge-and-Valley Appalachians, where Grandpop labored in a Bethlehem Steel limestone quarry. When he lost his job, Grandpop moved the family to Camden, New Jersey, and then to an abandoned stone farmhouse at a Malvern-area crossroads called Valley Store. The house had no running water or electricity, but Grandpop paid $17 to have a power line run to the kitchen. There was a well for drinking water and one to catch rainwater for washing clothes. The family would spend eight years there and take in friends from Camden as paying boarders. Grandpop and his two oldest boys worked in a quarry, shoveling stone, running crushers and planting dynamite.

For entertainment, Grandpop played the concertina, twanged a Jew's-harp and told stories for hours at night as the family sat in a room dimly lit by a kerosene lamp atop the fireplace mantel. The kids heard about King Arthur and his Knights of the Roundtable and Diogenes' quest for an honest man. In the kitchen, the oven warmed sweet potatoes or chestnuts for a bedtime treat. During the day, Grandpop hosted bocce games on the dirt front yard with iron balls from the quarry. Winners took home a bottle of homemade wine.

My grandparents moved their family for the last time to the house on East King Street in Malvern. Louie had quit school after eighth grade. In his teens, he caroused with his buddies, chased girls and learned how to fix up houses and repair vehicles. He was still a teenager when he changed his surname to Venditti, as an older brother had done after his well-traveled wife insisted that was how the family spelled it

My dad, Carmine Venditta (left), and his older brother, Louie Venditti, with their mother, Mary Venditta. The photograph was taken in the early 1940s in Malvern. Both my dad and Louie went on to serve in World War II. Dad was a Coast Guard radio operator on patrol frigates in the North Atlantic. Louie drove trucks for the 8th Air Force in England (courtesy the collection of Elizabeth and Carmine Venditta).

in the Old Country. She was right, but the rest of the family paid no heed.

Louie's World War II job driving fire trucks for the 479th Fighter Group in England included one experience that affected him deeply. The pilot of a crippled P-38 Lightning, returning from a mission over Nazi-occupied Europe, crash-landed on the base at Wattisham. His twin-engine fighter overturned and burst into flames, becoming a fireball by the time emergency crews reached it. Looking on helplessly, Louie saw the pilot upside down in the cockpit, banging his fist against the bubble canopy in a vain effort to escape. "That really got me," Louie told me. "It got me for a long time."

Back home after the war, he met fifteen-year-old Sally Caroline Gable while working as a handyman at her aunt's house in West Chester. The next year, she got pregnant. They were married August 12, 1948, when she was six months along. The plain civil ceremony took place in Elkton, Maryland, "The Marriage Capital of the World," where couples could get a license without having to wait for a blood test.

Nicholas Louis Venditti was born November 26, 1948. Years later, Louie and Sally had another child.

Louie bought a nineteenth-century wood-frame house down the street from his parents' place with the help of his employer, Foote Mineral, which made lithium chemicals and processed ores. He had started with the company as a furnace operator in 1942, the year the plant opened and before he got into the Army Air Forces.

The marriage was troubled from the start. The couple argued often. Louie spent much of his time at the VFW post, the American Legion and elsewhere. During a snowstorm early in 1957, he holed up for the entire night in Lou Thomas' taproom, and then all of the next day. Family lore held that he could tell barroom jokes for ten hours without running out of material. His mother called him a stinker.

Nicky was a second-grader at the public grade school when his parents broke up. Louie moved out, leaving the boy with his mother, and later filed for divorce. He met Bert at a local restaurant and bar where he would stop for a few beers on his way home from Foote Mineral. One night Bert was there with friends after roller-skating. She was divorced; Louie was still married to Sally. Bert liked him because he was polite, easygoing and personable. She would see him a few more times at the bar. It seemed that every time she was there, he was there, and they would talk. A few years later, in 1959, they married.

Sitting beside me at the kitchen table, Bert fell into sudden melancholy. Her head was bowed, eyes fixed on her lap, hands resting on her slacks.

"The thing I remember most about Nicky is his kindness," she said. "There was a little girl, Rosemary, who was slow and not pretty. Other kids teased Nicky about her in grade school because he was nice to her. He really felt sorry for her, and he used to feel bad. Everybody used to say, 'He's going down to see Rosemary.'"

She paused. "He was always teasing me, always trying to make me laugh."

She tried to say something more about Nicky, about how much she still missed him after all this time, but the words dissolved into sobs.

"Maybe this isn't a good idea," I said when her crying diminished to sniffles.

"No, it's all right, it's all right."

And she went on about Nicky as if nothing had interrupted her.

"He was a ladies' man, but he was nice. He was handsome, and he was a good kid, and the girls liked him. Oh, in the morning after a night out, he'd get me up, all excited, and say, 'I got coffee on. C'mon, I'll tell you what we did!' He'd tell me about somebody he'd been with and what happened. I won't tell you what he said!

"At that picnic we had for him, he tried to make a date with Vicky— she was the daughter of my friends Daisy and Frank Panic—and Terri was right there. Vicky kept telling him, 'But your girlfriend's over there!'"

Aunt Bert laughed.

"I don't think Terri knew he was going out with other girls. But you know, he loved Terri, even though he was skipping out here and there—and what kid didn't."

My aunt picked up a hardback book that was like a high school yearbook. The cover said: United States Army Training Center, Infantry, Fort

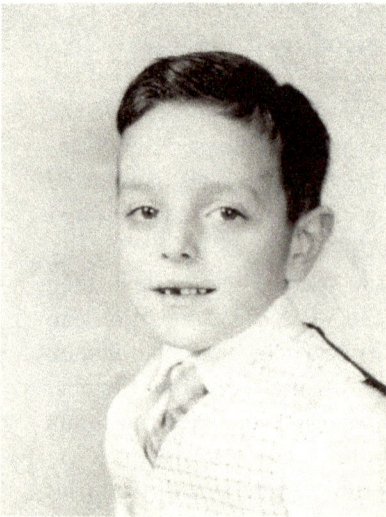

Nicky Venditti in the early 1950s. His parents, Louie and Sally Venditti, separated when he was in second grade at the Malvern Public School. He continued to live with his mother in the first block of East King Street in Malvern, near the old Pennsylvania Railroad tracks (courtesy the collection of Bertha and Louis Venditti).

Polk, Louisiana, Company C, 2nd Battalion, 2nd Training Brigade. It had pictures of the boot camp Nicky attended in the summer of 1968, showing recruits doing pushups, marching, practicing first aid, eating field chow, taking rifles apart, tossing grenades and pummeling each other with pugil sticks. There are black-and-white portraits of everyone in the class. Nicky's is on the bottom row of the second-to-last page, his face expressionless under the brim of his dress hat. Beside him is his friend Jerald A. Viall, and one row up is W.J. Vachon III.

"He was killed by the grenade, too," Aunt Bert said, pointing to Vachon.

She handed the boot camp book to me, got up and took a china cup off a shelf. She pulled a slip of paper out of the cup, which had been a gift to her from Louie. The paper was a note left with flowers on Nicky's grave on July 15, 1994, the twenty-fifth anniversary of his death. Uncle Louie would go to the cemetery almost every Sunday to put fresh flowers on Nicky's grave. It was on one of those visits that he picked up the note and brought it home. It said: "Well Nick, it's been 25 years. I still miss you. Love, Mary Anne."

I resolved to find her.

When it was time to go, Louie and Bert said I could take anything I wanted. I stashed the boot camp book, the letters, the telegrams and Nicky's certificates in my satchel. But there was one more thing Aunt Bert wanted me to see.

She had gotten Nicky a Catholic Bible for his twelfth birthday. Before he left for Vietnam, he opened it to the Book of Psalms and underlined the title "Psalm 23: The Lord, Shepherd and Host" in red pencil. He left the Bible on his dresser, open to the comforting words:

> The Lord is my shepherd; I shall not want.
> In verdant pastures he gives me repose;
> Beside restful waters he leads me;
> he refreshes my soul.
> He guides me in right paths
> for his name's sake.
> Even though I walk in the dark valley
> I fear no evil; for you are at my side
> With your rod and your staff
> That give me courage....

Bert and Louie didn't notice the Bible until after Nicky had gone. They left it open and undisturbed on Nicky's dresser for several years after his death, clinging to his parting act and the reassuring words he had left behind. Eventually Bert closed the book, but only after the exposed pages had become browned and brittle. Now she kept it in her room, next to a jar of pennies that Nicky had saved.

At times during my visit, when Bert was crying or there were long pauses of silence, I tried to think of something to tell my aunt and uncle that might lessen the sadness cloaking them. I groped for words and phrases—some were as simple as "I'm sorry"—but when I turned them over in my mind, all fell away as inadequate. Despite my sincerity, there was nothing I could offer. I kept quiet. Nor did I move to touch them. They might have thought I was signaling: Stop this aching in your heart. It was not a message I had a right to send.

Aunt Bert and Uncle Louie were inextricably connected to Nicky. He was part of their days, months, years. From him they drew a measure of happiness, devotion and fulfillment. But then violent, haphazard death had cut into that sublime state. They had gone on with their lives carrying a burden of unending sorrow. Ultimately, they would breathe their last and he would not be at their sides. On this day as I faced them, coaxing their memories of him to the surface, I felt the full force of his loss for the first time.

If anything could assuage Bert's grief, it was her belief in a theory that made Nicky's death easier to accept: It wasn't an accident that killed him, as the Army had said, but sabotage. "That's what happened, I think," she said.

I would discover later that Bert was not the only one who clung to an idea more bearable than what the Army concluded.

Louie offered no opinion that day I visited his home. In fact, he said nothing about his son's brief time in Vietnam. I tried not to push too hard in this first conversation, because I planned to return.

Two months later when I called to speak with him and Aunt Bert again, hoping to set up a second interview, Louie answered the phone. "I'm going in the hospital tomorrow to get my arteries cleaned out," he told me, as if it were nothing more than a routine physical exam. For much of his seventy-two years, he had disregarded his health. He smoked and drank too much. Recently, he had lost a toe to diabetes.

While he was at Bryn Mawr Hospital, his heart began to fail and he slipped into unconsciousness. When Mom called and said he was on life support, I drove from Allentown to see him. He was alone in the intensive care unit, flat on his back on a metal table in the middle of a room, arms at his sides and a hospital gown fitted around him. His eyes were closed, his face pale. Tubes protruded from his mouth, legs and forearms, trickling life into him from a machine.

"Are you his son?" a nurse asked.

"No, a nephew."

Uncle Louie would have given anything if I were Nicky watching over him. That was the way it should have been.

He never emerged from his coma. On February 18, 1996, he died of congestive heart failure. Father and son were reunited when Louie was buried at the foot of Nicky's grave, small American flags planted beside both of them. Louie's bronze VA plate says, "Louis C. Venditti, U.S. Army Air Corps, World War II."

He had outlived his son by twenty-seven years.

Chapter 4

Nicky Emerges,
Spring 1966

On one of a handful of days left in his high school career, Nicky and his friend Charley Boehmler were walking through the lobby of Great Valley High School when a large wall map of the United States caught Nicky's eye. Scattered across its face were pictures of many of his 288 senior classmates, each placed where he or she had been accepted to college, business school or a military academy.

"Look at this! Look at this!" Nicky exploded.

His friend was baffled. "What?"

To the side of the map, labeled "Work," were pictures of students who had no particular post-graduation plans. Nicky's was there.

Confronted with this withering reminder of his uncertain future, he took it like a wallop to the gut. So what if he'd been a less-than-attentive student who slid away for stolen days at every opportunity and did the minimum to get by. That didn't mean he could accept being a footnote to the accomplishments of his fellow seniors.

He knew he was really good at certain things. On the baseball diamond he had been a standout pitcher with a daunting fastball. He knew how to hunt and was expert with guns. He could coax a car to optimal performance and he had a decent singing voice. Yet standing in front of that map with seventeen years behind him and the rest of his life to go, Nicky sensed he was close to sinking into the oblivion of a dead-end, low-skills job.

This was the cousin I barely knew, yet now was compelled to know. I had to appreciate who he had been to put some measure of value on his short life. To do that, I had to find those who had seen him every day or at least had known him, those tightly connected to him or mere acquaintances. I would seek out their stories and their memories to allow me to re-create his life. It would give me what I never had before: an enduring picture of Nicky to help me grasp the meaning of his loss.

By the time he was about to graduate from Great Valley, Nicky had been meandering most of his life. His parents divorced when he was in elementary school. He'd lived ever since with his mom, Sally, and her gentle second husband on Malvern's main street in an unkempt nineteenth century house where she kept a twenty-pound raccoon indoors as a pet. Tension often hung in the air. His mother had a habit of yelling. She also could be less than truthful. Nicky learned early just what to do when bill collectors came. Without a word, his mother would run upstairs while Nicky answered the door and told the collector that she wasn't home and he didn't know when she would return. The man would frown, turn and walk down the porch steps and disappear down the sidewalk.

Louie, Nicky's dad, was much more fun. He, his second wife, Bert, and her son, Joe, lived right down the street. Louie was handy for teaching Nicky and the other kids how to play poker and drive his Jeep. But he could be surly after he had been drinking, and he was spotty about keeping promises. He'd tell Nicky he was going to take him fishing or whatever, then not show up. Nicky would sit for hours on his mother's porch waiting. Relatives down the street would offer to take him to a movie or out for ice cream, but he would say, "No, my daddy's coming to take me out."

Nicky eventually learned where to find his dad if he needed him. He'd go up the street to Lou Thomas' bar, where he'd knock and ask for Louie, who would come to the door, reach in his pocket and give his son a dollar or two for candy or baseball cards.

So the boy was on his own most of the time. He hungered for attention so strongly that he found himself a second father. Josiah Hibberd, principal at the old brick Malvern Public School, saw something in the wiry kid. He made lessons a treat, indulged Nicky's interest in sports and had one-on-one time with him, once having Nicky help him clean the cage of the eight-foot python kept in the school. Hibberd took Nicky and his forty-two classmates on dozens of day trips to places like Valley Forge, Gettysburg and the Jersey Shore.

At a solid six-foot-four, Hibberd commanded respect. He had the kids playing basketball, football, soccer and baseball. If the games were out of town, he packed the bats and balls and other gear into his blue-and-white Plymouth station wagon and he and some of the parents would drive the boys and girls wherever they had to go. Afterward he took his charges out for an ice cream treat at Mary Wilson's grille or the Guernsey Cow ice cream parlor and paid for it out of his own pocket. Hibberd made such an impression that even after moving on to junior high, Nicky returned to play sports after school and on weekends. Years later, as a young soldier going off to war in Vietnam, he would visit Hibberd to say goodbye and confess what was in his heart, something he could not bring himself to tell his parents.

But Nicky did not spend all his time searching for benevolent authority figures. He also acquired a fierce love of doing what he wanted to do, when and how he wanted to do it. All of Malvern was his playground, a grand stage for unfettered fun. The day after a winter's first snowfall, he and his pals threw snowballs from his driveway onto East King Street, lobbing them in a high arc so they would land on the roofs of passing cars. A spotter stood on the sidewalk and yelled "Now!" The police chief, Bill Cockerham, put an end to it after a volley went through an old woman's open window as she drove by. Cockerham punished the boys by making them throw snowballs at a brick wall for an hour.

Nicky ran amok with peashooters. He and some other boys bought bags of peas at Vince Carlino's grocery and ran up and down King Street shooting people. After they had made half a dozen trips to the store, Vince cut off their ammunition—he had followed them and seen what they were up to.

Before Nicky even reached his teens, he would make water balloons, go to the second floor of his house and roll them out the window onto kids passing by on the sidewalk below. He'd shut the window and laugh. It was just the sort of prank a kid from the poor side of town might pull—and Nicky was on the poor side, according to how the kids had mapped out Malvern. Warren Avenue, perpendicular to King Street, was the divider. West of it was the upper-class side; east was the lower-class side. If you lived north of King, on the other side of the railroad tracks, you were even poorer. Nicky's stepbrother, Joe, lived solidly in the poor section, and his friend Charley Boehmler lived safely on the rich side. Nicky was close to the line—just two houses east of Warren, but still with the poor folks. His pals needled him, "You're not on the upper side!"

King Street, the main drag, had Joe Cusano's barbershop, Mary Wilson's grille, Fisher's feed mill, Quann's hardware, Carlino's grocery, Buffington's garage, plastic cup maker Plastomatic, Lou Thomas' taproom, as well as Grandmom and Grandpop's house. For years, my grandparents and their pack of children, including my dad, had been intimately connected to this community, sharing its joys and mourning its tragedies. When two boys drowned in the pond at Malvern Preparatory School in 1949, my Uncle Sam was among the volunteer firefighters who searched the water for the bodies as my parents stood in the horrified crowd looking on.

Malvern still has the feel of a small town, tucked away from the mainstream, an idyllic throwback. Settled by Quakers in the nineteenth century, it grew up around a railroad junction built on a ridge and stayed true to its humble origins. Philadelphia's westward spread, which has turned much of the rest of Chester County's rolling hills into townhouses and strip malls, has not overwhelmed Malvern. It remains a charming place of neatly arrayed frame houses, stuccos, brick Victorians, Edwardian twins. In its heart, St. Patrick's Catholic Church and the First Baptist Church stand side by side. The outside world partly intrudes in the form of the railroad tracks, which carry Amtrak trains and a commuter line of the Southeastern Pennsylvania Transportation Authority through the borough.

But this peaceful little town is nonetheless far different from what it was in Nicky's day, exuding a new air of sophistication. Quaint shops lining King Street sell fine art, equestrian supplies, decorative hardware. There are salons, boutiques, a music school, a yoga studio. Redevelopment on East King has produced Eastside Flats, a pair of nearly block-long, three-story-high structures offering stores, restaurants and luxury apartments. Lou Thomas' old bar is now the Flying Pig Saloon, with craft beers on tap.

In the Malvern of the 1950s, Nicky found his summertime passion—Little League baseball—on a field a little more than two blocks west of the churches. He started playing when he was in fourth grade as a pitcher and infielder. The diamond turned out to be his rightful place, a showcase for his natural athletic ability and single-minded focus. He became a Little League star, resolving his need for respect and recognition on the pitcher's mound. His curveball was good, but it was his wicked fastball that baffled the boys at the plate, earning him an all-star berth in the Chester Valley league.

Still, Nicky's commitment to baseball didn't cap his exuberance or fill the void in his pre-teen life. There was a place that would go a long way toward making up for that, a wild, open part of town where a boy could go and be anything his imagination would allow.

To this day, townsfolk call it the Waterworks, forty-one acres of dense woods plus a few acres of open ground on Malvern's eastern edge, where the borough had a well and a pump house. Louie and Bert bought a gray, two-story frame house across the street when they married in 1959. The spot was ideal for Bert's son, Joe. Nicky, just blocks away, could walk or ride a bicycle there in a few minutes. It would become so central to the boys' lives that when Louie talked about moving several miles away, Nicky and Joe protested: Joe wouldn't have any way of getting to the Waterworks. Louie backed off. He promised that he and Bert wouldn't move out of town until after the boys got their driver's licenses.

For rambunctious kids, it was liberating to go effortlessly from sidewalks and streets into the dark, cool, sloping woods where few others ventured. The site is much the same today—though the well has long since been filled in and the pump house is only used for storage. It is easy to understand what drew the boys all those years ago. Towering pines and leafy trees allow only slanting shards of sunlight to creep through vast, green canopies overhead. Deadwood and tangled undergrowth cover the ground. Fallen, cracked trees form a crazy architecture, making for handy hideaways. Thickets encroach on winding footpaths, in some places forcing a hiker to duck to go on. And aside from the trickle of water in a brook nearby, utter silence blankets the mystery of the place.

Nicky and his friends could really cut loose in those woods. They almost lived there. They'd spend all day playing army, building forts, using sticks for guns and rocks for grenades. They'd camp out at night, snag frogs in the creek, play hide 'n' seek and tag, fire BB guns. When they were older, they sneaked smokes.

Along the woods lie rolling fields, steep enough to sled down when snow-covered. In warmer weather, the boys brought their gloves and a ball for a catch. Nicky and Joe learned to drive on the looping, gravel access road. At a police firing range on the site, with Chief Cockerham's permission, they learned to shoot. They hunted groundhogs and pheasants and took aim with their .22 rifles at blackbirds perched on power lines. Directly across from Louie and Bert's house, just beyond a grove of walnut trees, a flat field stretched to the woods—broad enough for baseball. Louie would cut the grass so the kids could play sandlot ball there, Wiffle ball, softball.

The Waterworks was Nicky's escape and the common ground where he and Joe became friends. Their relationship as stepbrothers had been rocky at the start. While Nicky had get-up-and-go and never seemed to tire, Joe couldn't keep up with him. Nicky picked at him, goaded him, but Joe couldn't come out on top in a tangle with a kid who had rock-like muscle tone. Nicky's muscles didn't pop; they were hard. He had raw power and toughness. Coming up the street one day, he went to shake hands with a friend who had just stepped out of his house, then flipped him over his head onto the ground—all in fun, which to Nicky meant pulling harmless pranks.

His visits to the firing range at the Waterworks hooked him on guns, seductive symbols of freedom and control. Target shooting and small-game hunting were his pursuits, but his interest went beyond those pastimes. He memorized ballistics charts so he could tell the trajectory and bullet energy of various calibers and powder loads. He studied weapons publications and decided on his own which guns and accessories to buy. He and Joe accumulated shotguns, pistols and rifles, firing on the Waterworks range and a police range near the Malvern

Nicky (left) with his stepbrother, Joe Gray. The boys hunted raccoons at night at Malvern's Waterworks or on the grounds of the Catholic retreat, St. Joseph's-in-the-Hills. When they were older, they raced their cars on Route 30 (courtesy the collection of Bertha and Louis Venditti).

train station. Outside town Nicky would hunt groundhogs on an uncle's property. When he got a fine .22 Magnum rifle, his friends envied him. Their envy grew when he got a Tasco scope while the other boys used open sights. Then Nicky got a slick .22-caliber Remington nylon-stock rifle while the others all had wooden stocks.

Small-game hunting in town got the boys some unwanted attention. It happened because Joe liked to hunt raccoons at night when they were out. It wasn't Nicky's favorite time to prowl the woods, but he could usually be persuaded, seeing the invitation as an opportunity to get out of the house and smoke, which his parents would not have allowed.

"How many cigarette butts can you find in all the ashtrays?" he asked when Joe called once. "If you find me enough, I'll be down."

In the darkness, they'd hunt raccoons at the Waterworks or jump a fence onto the adjoining grounds of the Catholic retreat, St. Joseph's-in-the-Hills. The caretaker got calls from folks who saw lights moving in the woods—the boys' flashlights—and the police chief got involved.

"Joe had this great big old coon dog out back, tied to a box," Chief Cockerham remembered many years later. "Big brown dog with big ears. When he gets on a coon, he chases him to a tree. He stands there and gnaws at the bark, *arrrooooom, arrrooooom, arrrrooooom*. Then here comes Nicky and Joe with the flashlights to spot the coon up in the tree with. They'd come out with these big old heavy stockings, old jackets. They were such hellish night boys, the both of them, and they paired up so nice that I took a liking to them. I called them my boys."

The chief got Nicky and Joe permission to hunt in the woods at night.

Gradually, the Waterworks was overtaken by Nicky's other passions. School was still his lowest priority, and he skipped out whenever he could. When he and Joe were going to General Wayne Junior High— named after Revolutionary War General Anthony Wayne, a native of the area—they'd meet at a bus stop between their houses. Whoever didn't want to go to school would wrestle and hold down the other until the bus left. That way, they both stayed out.

Later at Great Valley High, Nicky and Joe would bag school to work on their cars. That's what they were doing one day at Louie and Bert's house outside of town when the phone rang. It was the school's disciplinarian, Albert Como, singer Perry Como's brother. Nicky answered, pretended to be Louie and said, "Yeah, Joe is sick," and hung up. He ran out of the house, telling Joe, "I gotta go," jumped in his car and reached his mom's house in time to answer Como's call about why Nicky wasn't in school. This time, Nicky was himself. "Oh, I don't feel so good," he said.

He got a kick out of doing the same kind of stunts in school, which he considered less a place for learning than for ogling girls and horsing around. He had a way of starting trouble while avoiding the consequences. In shop class once, he planed boards so the sawdust blew into the faces of two friends, and when they moved to retaliate, the teacher caught them, sending them out to run around the soccer field as punishment. Nicky would wrestle and hold a pal in the hallway to make him late for class, while his own class was close enough for him to make the bell.

He had always been charged up about girls. When he was twelve, he stashed girlie pictures in his room. He had a brother called L.B., who was ten years younger, and Nicky in his early teens romanced the little boy's babysitter while she was supposed to be working. But it was hard for him to get dates in high school, in part because of a severe acne problem. His friends teased him, calling him Picklenose, which he didn't like. He was sensitive about it, and frustrated because he went for the pretty girls in a big way and wasn't shy around them.

Karen Armstrong was one of the girls Nicky was keen on. She was queen of the top dance of the year, the Sports Regalia, and had biology lab with him. One day in the lab, he went up to Karen and, in an effort to create a see-through effect, squirted the front of her blouse with distilled water.

Nicky had his best luck with younger, more impressionable girls. He once took the sister of three buddies for a ride, then stopped the car and surprised her with a passionate kiss. Patty Forrester was in her early teens and good-looking, and her brothers were fiercely protective. Though she liked that Nicky had come on to her, she didn't want them to know. She was afraid of how they would react.

"That one kiss was my Kodak moment in my brain, and I'll never forget it as long as I live," Patty said years later. "I wonder where it might have gone, if he had shown more of an interest, or if he had said, 'Let's go out.' He always seemed sure-footed but contemplative, with a calming quiet. I can see him never saying a word about that kiss because of this quiet man that he was."

Nicky would not have a steady girlfriend until he met Terri Pezick after graduating, but he did have a close friendship with one girl that would continue until his death.

He met Mary Anne Wallace one Saturday night about 1964 at a fire company dance. They never went on a date; they just talked. Day or night, Nicky would show up at her family's rancher. Sometimes they would sit on the front porch with her father, a mechanic who once ran a garage in town—and whose son, Mary Anne's older brother, was one of the boys who drowned in the Malvern Prep pond. Nicky liked to gab with him about cars. He had a '62 Chevrolet convertible that he kept waxed, shiny and spotless—cleaner than most people's homes. With gasoline cheap at forty-one cents a gallon, he and Mary Anne would drive around, to nowhere in particular, for hours. She found she could talk to Nicky about her feelings. He was a good listener, more mature than other teens, and seemed to have an inner toughness that had helped him take his parents' bitter divorce in stride. He knew when it was time to listen, when to make a suggestion.

"He was also out for fun and partying," Mary Anne remembered. "He was a prankster. He would start something and then fade from the scene. Everybody else would get in trouble for it. He enjoyed that role. It was all good, clean fun. He was out to have the best time he could have without hurting anyone."

Though she was interested in him, she saw herself as quiet, immature and uncertain how to handle a relationship. She couldn't tell whether Nicky sensed there was a chance for them to get closer. If she hadn't been shy, she would have told him how much she wanted to be with him.

She would never have that opportunity. Nicky met fifteen-year-old Terri Pezick at the Sinclair service station on West King Street where he pumped gasoline. Her mother did inspection reports and billings for the station manager. She thought Nicky, who was seventeen and now graduated from Great Valley High, was a nice young man and introduced him to her daughter. They would spend so much time together that Terri would have to repeat eleventh grade. She flunked the year after missing too many school days.

It hadn't mattered to her whether she went to classes or not. Nicky was more important to her. He would pick her up at the bus stop or at her home in the morning and whisk her away in his Chevy convertible. They'd drive around, browse the big mall at King of Prussia or hang out with friends.

She was pleased to be dating a guy who was more than two years older than she, and looked up to him and allowed him to influence her. Nicky said he didn't like her name ending in a "y" because it looked like a guy's spelling, so he changed it to the feminine Terri, and she spelled it that way from then on. He wanted her to be ladylike, once dumping her from his lap because she said "fuck."

Nicky's friendship with Mary Anne proved to be a regular sore spot. Terri was convinced that Mary Anne was trying to steal him away from her. She got so furious she called Mary Anne and said, "Would you please get your own boyfriend and leave mine alone!"

Though disappointed and puzzled by Nicky's choice, Mary Anne

Terri Pezick poses with her poodle, Claudette, a gift from Nicky. Terri and Nicky became engaged at Christmas 1968, when he was home on holiday leave from the U.S. Army Primary Helicopter School at Fort Wolters, Texas. Nicky kept the photograph in his wallet and took it with him to Vietnam (courtesy the collection of Bertha and Louis Venditti).

continued to see him casually. Terri gave him grief about that and bristled whenever she saw them together, even if they met by chance. During one of Nicky's visits to Mary Anne's home, he told her, "I'm putting another nail in my coffin by being here." The two girls would never get along. But they would call a truce for Nicky's sake on the day they and the rest of Malvern heard what happened to him in Vietnam.

On a par with Nicky's affection for girls was his love affair with cars. He had an instant connection with them, just as he'd had with baseball and guns. He saw them as tantalizing mechanical wonders to be understood, finely tuned and run to the limit of their capabilities. He got a thrill from racing on the street, how wild it was, the exhilarating rush of speed and power. He liked how the coolness of driving a hot car got people's attention, lured women.

Nicky's keen interest in wheels was a major reason he only did enough school work to get by. As soon as he could, he got a job at a gas station and worked many hours on nights and weekends to earn money to buy a car. He had played baseball for Great Valley High but quit by his senior year so he could use the time to work in support of his pastime. His co-worker at Sam Wasson's Sinclair station, Charlie Thomas, was a mechanic and Air Force veteran of the Korean War who found Nicky to be a quiet, bright kid who was always laughing. At the time, Nicky had a beat-up green 1958 Chevy sedan. They would race two or three times a night after closing the station at nine.

With a mind to go faster, Nicky saved his money and got his "hot car," a 1962 Chevrolet Impala, a maroon two-door convertible with a white vinyl top. He had bought it off a used-car lot during his senior year at Great Valley. The car was big, heavy and quick off the line, with a hepped-up ignition system, a 327-cubic-inch V8 engine with 300 horsepower and a Hurst four-speed floor shifter. "Nicky's Chevy was the hottest car around. Man, it could rip!" a cousin remembered. Charley Boehmler called it a babe-catcher.

Nicky quickly learned how to coddle his car. Charlie Thomas could show him something one time, and he picked it up. If he was going to race his Chevy, he knew to dump a mothball into the gas tank—the methane in the mothball would raise the octane level in the gasoline, and the car would run a little faster. He and Charlie would run the Chevy east on four-lane Route 30 to the village of Wayne. Sometimes he'd take it to Upper Darby to the Hot Shoppe, a drive-in burger joint that was a racers' hangout. They'd go looking for someone to race. It was the thing to do on the Main Line.

At first, Charlie did the driving because Nicky felt he couldn't shift fast enough. Nicky would laugh throughout the race. Several times, cops stopped them and gave them a warning. The two would go back to Malvern and Nicky would tell his pals how his car performed. If something broke, he and Charlie would work on it in the shop until two or three in the morning, until they got it purring again. Nicky would work on the car that late even on school nights.

When he learned to shift fast, he took the wheel himself. To win on the usual quarter-mile stretch, at speeds of 100 mph and beyond, he had to take consummate care of his car, drive skillfully and keep cool under pressure. It raised Nicky's competitive spirit. As a race was about to start, he would be transformed as determination took over. He'd focus and grit his teeth. If he lost, he would brood about it and ask himself: What can I do to make this car faster?

Racing on the streets was a subculture with its own unspoken rules of engagement played out at places like the Hot Shoppe in Upper Darby and the Gino's fast-food restaurant in West Chester. Charley Boehmler remembered what it was like to walk into Gino's with Nicky: "Nicky would go in there looking for someone to race. People would be looking for other people to race, and of course, Nicky was always looking for girls, always trying to scope something out with his car, and we'd end up with somebody who wanted to race. There'd be nothing said. Somebody would look at you, and it was a given. It would be set up without a word. They would both leave at the same time and go down to the stretch."

Silent set-ups also were made on the street. Strangers would connect and jockey until they both came up to a light. They'd slow way down, side by side. The light would turn red. When it turned green, they'd be off. It went on two or three nights a week throughout the summer for Nicky and Joe, who had his own car to race, a 1966 Ford Fairlane GT.

Cars took a central role in Nicky's pranks, which expanded to include what Joe called car decorating episodes. Nicky would get Charley Boehmler to help him soap the windows on Joe's car or smother it with shaving cream. Then Nicky would join Joe in decorating Charley's car. One day, Joe found his car with an old plastic swimming pool over the top, a "for sale" sign on the hood, a lady's coat hanging from the antenna, toilet paper wrapped all around, soap on the windows, shaving cream down the side. Nicky and Charley had created this beauty. Now Nicky went to Joe and said, "We're gonna get Charley for this!" So the way Nicky played it, his car never got the treatment; the other guys had the mess.

As the high school days of pranks, hunting and target shooting, chasing girls and racing on the streets waned, Nicky sensed he was careening toward the time when he would leave Great Valley High for the adult world—a place for which he was not prepared.

A friend asked him what he planned to do after graduation.

"Well, I don't know."

"Nicky, the rate we're going and the rate the country's going, I won't have to worry about it because I'll be drafted," the friend said.

"They're not going to draft me," Nicky said. "I'll probably enlist because I don't want them telling me what I'm going to do."

With graduation day nearing for the Class of '66, other young Americans were sweating out a widening war in Vietnam. U.S. ground combat troops had been there for fifteen months, and U.S. aircraft had been bombing North Vietnam for longer than that. By the end of the year, American troop strength reached 385,000, more than double what it was at the end of 1965. Eight thousand, four hundred and seven Americans had died, more than two-thirds of them in 1966 alone.

Great Valley's Patriots, as its sports teams were called, had been toddlers the last time their country had fought a war, against the communists in Korea. Now, with the United States taking on another communist threat in the same part of the world, the students wanted to do something for the troops. They held a drive to bake, package and ship thousands of cookies. Thirty pictures in the yearbook show the kids mixing batter and tending to stoves in the school cafeteria, and GIs and South Vietnamese civilians unloading cardboard boxes from Air Force cargo planes and piling them onto flatbed trucks.

"This was a year to remember," says the yearbook, Musket 66. "Batman flew onto the television screens as the Cookies for Vietnam were flown out.... Everyone in student council made a mad dash for the Cookie Control Center to wrap, pack, label and eat the cookies going to Vietnam."

Elsewhere in the book are the portraits of the graduating seniors. Under each is an expression chosen by the editors from a thick book of quotations. They would try for literary lines that touched on some aspect of the student's personality or character, or played on the person's name.

Under the photo of a smiling Nicholas L. Venditti, with a mention of his only activity, baseball, was this line attributed to Anonymous: "The quiet man who rises to his cause."

On the day Nicky and Charley Boehmler stood looking at the hallway map pointing to the seniors' futures, as Nicky felt rage and embarrassment because his photo was off to the side, he made up his mind that he was going to find some avenue to success, some way to lift himself above mediocrity.

It would not happen right away. He would continue pumping gas, racing his car, hunting and target shooting, instigating pranks, doing as he pleased. In the fall, when he turned eighteen, he would move in with his dad, stepmother and Joe at their new home, a rancher outside of town. After a year, he would get a job with Joe running forklifts in the shipping department at Plastomatic in Malvern—unskilled work, the only kind he was qualified for. He and Joe, assigned to the overnight

Nicky's Great Valley High School graduation photograph. The yearbook, Musket 66, lists his only activity, baseball, and a line attributed to Anonymous: "The quiet man who rises to his cause."

shift, would make the plant a new venue for mayhem—shooting water guns at the operators, tossing plastic cups at each other, slathering red and black ink on machines and phone receivers as gags against the unsuspecting.

But a seed had been planted in him, a kernel of intent.

Chapter 5

Best Friends,
October 24, 1996

Tony Viall peered out his kitchen window over the sink, eyes frozen in a squint, and dragged deeply on a Winston. He blew out the smoke in an easy stream. It drifted in swirls about his ruddy face and thinning, reddish hair. On this sunny day in October 1996, he was remembering, at my request, events almost three decades old. He was remembering my cousin Nicky.

Still lean at forty-seven and hardly showing the effects of wounds he had brought home from Vietnam, Tony had arranged to take a few days off from his job at a chemical plant to tell me how he came to know Nicky. We spoke at his home in Ooltewah, Tennessee, twenty miles from Chattanooga.

"The beginning was Fort Polk. Nine weeks. Nick and I were in the same company, the same platoon, the same squad. We bunked next to each other. In formations, we lined up next to each other.

"We were away from home," Tony said in a drawl as soft and slow as the smoke. "It was a restricted environment. You didn't have a regular social life with people. Your privacy was stripped away. You'd eat and sleep together every day. You were with each other twenty-four hours a day, so you'd either become friends or enemies."

They became friends, bonded strongly by ordinary passions like girls and cars, sports and beer-and-pizza good times. Both children of broken families, both seeking a decent future in the skills they would be taught by the military, they were a natural fit. But they almost did not cross paths. In what would have been a cruel setback in his push to rise above the mediocrity the high school map had pinned on him, Nicky's dream of being an Army pilot was nearly crushed at the start.

Before signing up, he had to take an aptitude test to determine whether he qualified for flight training. The written test he took December 19, 1967, at the Armed Forces Examining and Entrance Station on North Broad Street in Philadelphia proved intimidating. At one point, would-be fliers had to determine the position of a plane in flight by looking at two dials, one showing the artificial horizon, the other showing the compass heading. Other parts tested an applicant's ability to judge distances and visualize motion, principles of helicopter flight and of mechanics in general.

Nicky failed and fell into a deep funk, but the demanding maw of Vietnam would resurrect his wish. With losses piling up, the Army needed more chopper pilots. He would get his chance to fly, his ticket to what he was sure would be a better life.

"The recruiter just called me!" he told his hometown friend Charley Boehmler. "They lowered the requirements! That means I qualify!"

Nicky passed the physicals, and on April 18, 1968, ten days after the end of the war's biggest battle to that point, the siege of Khe Sanh, he enlisted to become an Army aviator. On June 25, a few weeks after Democratic presidential contender Robert F. Kennedy, whose peace platform had just won him the California primary, was gunned down on the campaign trail, Nicky reported for duty. He was off to Fort Polk, taking the first airplane ride of his life to spend the summer in several hundred square miles of hot, humid, swampy western Louisiana. There, he found friendship with Tony and others, including Billy Vachon, an ex–high school star athlete from Portland, Maine, the recently married father of a baby girl.

Early in World War II, the Army had built what was then Camp Polk and used it to train millions of fighting men. It was modernized in the 1960s as an infantry training center. By the time Nicky, Tony and Billy arrived, it was providing more infantry replacements to Vietnam than any other U.S. training center.

Together the buddies sweated out boot camp's rigors—the drill instructors, the daily physical regimen, Army indoctrination. A photo in their Company C graduation book shows Nicky, Billy and a couple of other tired recruits on the ground in a stand of trees, eating rations. Nicky is in the foreground lying on his side, propped up on his right elbow, looking over his shoulder. His hair is cut down to fuzz, his face a little washed out by the camera's flash. Tony is with them, but only his stretched-out legs are in the picture.

"In the evenings when you finished drilling," said Tony, who grew

Nicky in 1968 as an Army infantry trainee at Fort Polk, Louisiana. On the back of this photograph, which he sent to his girlfriend, Terri Pezick, he wrote: "Charlie Company. Vietnam or Bust. July 14, 1968. To Terri with all my love. Nicky."

up in Rossville, Georgia, in the Peach State's northwestern corner, "you had to do things like shine boots. So you're sitting there on your bed or on your locker and you're talking. We talked about cars and women, things like that."

Nicky told him about his girl, Terri Pezick, back home in Malvern. He'd be thinking of her when he sang a few measures of the Temptations' 1964 hit "My Girl."

Tony knew she was special to him; Nicky made it clear there was no one else he wanted. A black-and-white snapshot Nicky sent Terri

Left: Nicky as an Army basic training graduate in August 1968 at Fort Polk, Louisiana. He was with Company C, 2nd Battalion, 2nd Training Brigade (U.S. Army). *Center:* Jerald A. "Tony" Viall in the Fort Polk graduation book. Tony was from Rossville, Georgia. He and Nicky became best friends in boot camp and went to flight school together (U.S. Army). *Right:* Wilbur J. "Billy" Vachon III as a graduate of the Fort Polk infantry training center. Billy, from Portland, Maine, had been a standout athlete in high school. He was married and the father of a baby girl (U.S. Army).

from Fort Polk shows that she was on his mind. Taken a week after training started, it shows him posing with one knee on the ground and a forearm resting on top of the other knee. He's scruffy in dusty fatigues and boots. The edge of his helmet is even with his brow, partly shading his eyes from the glaring sun. With his jaw set, he looks tough and ready to fight. On the back he's written: "Charlie Company. Vietnam or Bust. July 14, 1968. To Terri with all my love. Nicky."

Company C of the 2nd Battalion, 2nd Training Brigade completed training at Fort Polk on August 30, the day after the Democratic National Convention in Chicago adjourned after nominating Vice President Hubert H. Humphrey for president. The convention had been a riot-filled fiasco. Antiwar demonstrations and violent clashes between police and protesters riveted attention on the nation's social conflicts and helped doom Humphrey's candidacy. The Republicans had met in Miami earlier in the month and nominated Richard M. Nixon for president. In his campaign for the general election in November, Nixon claimed he had a secret plan to end the war, but he did not. Thousands of troops kept flowing into the conflict, including most of the 218 young men in Company C.

With basic training over, Nicky, Tony and Billy rode a chartered bus from the flatlands of Fort Polk to hilly north Texas, about eighty miles west of Dallas. They came to the modest town of Mineral Wells, dubbed Miserable Gulch by the aviator candidates for its paltry entertainment and dearth of young people other than soldiers. Here, at U.S. Army Primary Helicopter School at Fort Wolters, they would learn to fly. It would be their home for the next twenty weeks.

Tony and Nicky were among 200 potential fliers who started the Warrant Officer Rotary Wing Aviator Course on September 3. They now were on the Army's assembly line for producing common-man fliers quickly and economically as no other branch of the service did. Here no West Point training or college degree was needed. The Army had realized in the 1960s that Vietnam demanded a steady flow of chopper pilots and it adjusted accordingly, turning out as many as 600 a month in a fraction of the time it took the Air Force, Marine Corps, Navy or Coast Guard to graduate fliers.

The men in Class 69-13 wore orange caps and were divided into four sections of fifty each. Nicky, taller than Tony, was put in a section destined to learn to fly in a bigger chopper. There was little association among the sections, so the buddies from boot camp weren't in close contact any longer. That annoyed Tony. "You made a friend you'd been with nine weeks," he said, "then you were among strangers again."

Fort Wolters had been training military pilots since 1956. It had three heliports and twenty-five stage fields, where student pilots learned to land in a confined area and pull out safely. But the first few weeks, Tony recalled, had little to do with flying and a lot to do with harassment.

"They'd wake you up in the middle of the night for fire drills. On Saturdays we'd double-time to places, pick up rocks and carry them back and put them in a drainage ditch. On Sundays we'd pick up the rocks and take them back. It was a weeding-out process."

The first month was called pre-flight and was limited to ground school. Typical subjects were weather, map reading, navigation, the mechanics of the aircraft and aerodynamics. It also included military classes that prepared the men for the officer corps.

Nicky found the weather instruction especially challenging. "I start flying in one week if I can cut the cake at school this week," he wrote to his mom, Sally Pusey. "We are having weather and it is very hard to learn. Last semester, half of the class failed it!" But he said the nuns at Immaculata College near Malvern were praying for him—he didn't say how he knew that—so he was sure he'd make it.

The remaining sixteen weeks at Wolters were the pre-solo and solo phases. Pre-solo involved riding with an instructor pilot. Nicky flew the Hiller OH-23 Raven, while Tony flew the Hughes TH-55 Osage. Both were light observation helicopters, no-frills aircraft with bubble cockpits.

In October 1968, Nicky sent Terri a series of photos of himself and OH-23s on the ground and in the air. On the back of one picture, he wrote, "Terri, this is what I've been flying. I finally managed to get a picture for you. I'm sorry it took so long. Also, I love you and I sure wish I could see you, angel. Take care. Remember I love you with all my heart. Love, Nicky."

Nothing about flying a helicopter was easy, but hovering, the maneuver that makes the craft unique, was particularly difficult for most students. They had to keep the craft stationary, three feet above the ground and pointing in one direction, which was like "balancing yourself on a big ball, talking, reading a book, eating and drinking at the same time," as one Wolters graduate described it.

A whimsical card issued to Nicky on October 23, 1968, designated him a "genuine hoverbug" for "having remained motionless in space; flown backward, forward, sideways and vertically in U.S. Army helicopters."

Then came the students' first solo flights, the apex of achievement for a student pilot. Nicky described the landmark experience, lasting only a few minutes, on the photos he sent Terri: The instructor got out of the aircraft and the student held it at a hover and approached a take-off lane. The instructor observed from the control tower and had radio contact with the student. The student got instructions to take the helicopter up and fly the traffic pattern—turn into the crosswind, then downwind, then crosswind again, then turn for the approach and land upwind.

Afterward Nicky and the others boarded a bus and stopped at the Mineral Wells Holiday Inn, which had two rotor blades set up in its yard in an upside down "V." A few years later, a sign would declare, "Under these rotor blades passed the finest helicopter students in the world. July 1, 1956–August 16, 1973." Now, in what had become a Wolters rite of passage, Nicky and the other guys who had just soloed for the first time were dragged through the arches by their fellow students and dumped into the motel's pool.

Having reached the level of solo proficiency, they learned to fly cross-country and at night—sometimes solo, other times with an instructor.

"Things started becoming more lax," Tony said. "We had passes for the weekends, and so we started spending some time together again. We might go to Dallas for a Saturday afternoon, come back Sunday."

But even a weekend pass wouldn't give Nicky enough time to go home and see Terri. Their long separation had made him yearn for her more intensely. He got a sharper idea of where he wanted their relationship to go, and he was impatient about getting there. He wrote to his mother, cluing her in:

"So Terri is being good. Well I sure don't get as many letters from her as I used to. That's the breaks! I'll take care of everything Christmas. I'm giving her an engagement ring Christmas, I think."

As the holiday season neared, Nicky and Tony did something impulsive that could have damaged their chances of getting their wings. In a bout of acute homesickness one weekend, they went absent without leave.

"There was a satellite exchange that had a little pizza parlor. We were drinking beer and eating pizza, and we decided we were going to go home," Tony said. "So we caught the shuttle bus to Mineral Wells and got bus tickets to Dallas. We were going to buy airplane tickets there. By the time we got to Dallas, we were sobering up and we realized we were probably doing the wrong thing." They scurried back to base in time to avoid detection.

Weeks later, Nicky went home to Malvern on a proper leave as Fort Wolters' flight school shut down during Christmas and New Year's. On Christmas Eve, he bought Terri a quarter-carat engagement ring. They picked it out together at a jewelry store in the King of Prussia Mall. Later in the evening they went to Mass.

Less than a month of training at Fort Wolters remained. Some of the student pilots had already washed out, unable or unwilling to continue, not up to snuff in academics, military development or flying ability. Health also came into play for some who might find, for example, that they couldn't overcome vertigo. But 173 of the original 200 in the 6th Warrant Officer Candidates Company graduated at the end of January 1969, each with about 110 hours of flying time, most of it solo. Among the successful students were Nicky, Tony and Billy.

In Tony's platoon was a college dropout from southern California who would go to Vietnam with them and be in the classroom at LZ Bayonet on a fateful July morning. His name was Tom Sled.

Glad to have finished their Fort Wolters training, Nicky, Tony and their friend Skip Smith—Tony's buddy from high school who joined

Edwin A. "Skip" Smith, Jr., as a Fort Polk infantry training graduate. He and Tony Viall went to high school together in Rossville, Georgia, and joined the Army to become pilots. After earning his wings, Skip went to Korea instead of Vietnam (U.S. Army).

the service with him—celebrated with about two dozen others in a barracks room drinking beer and eating pizza. So many guys in one room violated the rules, and when a tactical officer showed up, he had them trapped. "Tac" officers were like drill sergeants and usually were Vietnam returnees.

The officer told all of the men to produce their demerit slips so he could fill them

Thomas Sled at a UH-1B Huey helicopter while training at the U.S. Army Aviation School at Fort Rucker in spring 1969. A helicopter ride at San Diego's Mission Bay had hooked him on flying. He was with Nicky, Billy Vachon and Tony Viall in the orientation building at LZ Bayonet when the explosion happened (courtesy Thomas Sled).

out, meaning they wouldn't get weekend passes. Nicky and Tony were close to the door and told the officer they didn't have their slips with them; the slips were in their rooms. The officer allowed them to leave to get them. A few others gave the same excuse, and he let them go too. But then he realized he'd been duped, and no one was coming back. Skip didn't make it out of the room and got slammed with losing a month's pay. Later as he, Nicky and Tony were driving to Rossville in a Chevelle that Skip had bought in Texas, his buddies teased him. "You dumb ass, you should've been near the door."

The next phase of training was Army Aviation School, sixteen weeks at Fort Rucker in Alabama, about eighty miles south of the city of Montgomery. Nicky and Tony were back together. They shared a room with Skip and had the same instructors.

Training for Class 69-13 began February 5. During the first four weeks, they learned the basics of instrumentation, followed by four weeks of advanced instrument training. That had them flying a small helicopter called the Bell TH-13T. They got a total of fifty hours of flying time in the instrument phase, which Tony and Nicky found to be the hardest part of flight school. One aspect of it involved riding with an instructor and wearing a hood over their heads like blinders on a horse. It prevented them from seeing anything but the instruments, their only way of knowing the helicopter's position in flight.

In letters home, Nicky once again fretted about whether he'd pass muster. He expressed his doubts—and homesickness—in a mid–February letter to our cousin Mike Beam, then a Marine reservist specializing in aircraft maintenance.

"When am I going to get my [warrant officer] bars? Well, I'll probably never get them. This stuff is so hard and I'm so dumb, I'll probably get kicked out.... Maybe one of these days I'll get home again. I almost forgot what it looks like! ... Well, it's 7 now. Bedtime in three hours, up at 4:30. At 6:45 I'm in the air, trying to fly those damn instruments. I doubt if I'll ever learn."

He revealed to his mother even deeper doubts that he apparently shared with no one else. The prospect of a three-year obligation coupled with a hazardous tour of duty in Vietnam depressed him. "The longer I'm here, the more I wonder whether I want to be an Army pilot or not. I don't think I'll get a kick out of signing up for three more years, plus flying around while someone shoots at me. But I went 24 weeks now, so I might as well finish."

Nicky hung on for the last eight weeks at Rucker, long enough to learn to fly an icon of the Vietnam War. That was the Huey, or UH-1, synonymous with the American presence in Vietnam and seen in such films as *Apocalypse Now*, *Platoon* and *We Were Soldiers*. Though the military had used choppers during the Korean War, their main role was plucking casualties from the battlefield. Vietnam changed that.

The Huey was fifty-seven feet long, the length of a Tyrannosaurus Rex or a humpback whale. It weighed more than two-and-a-quarter tons, could cruise up to 125 miles per hour and had a range of around 300 miles. It landed on a pair of skids and was strong enough to carry more than a ton of cargo. Metal blades made up its main and tail rotors, the semi-rigid one above drooping like antennae when it wasn't spinning. Bell Aircraft Corp. developed the Huey because the Army wanted a general-purpose helicopter. Flown for the first time in the fall of 1956, it was the service's first aircraft with a turbine engine, a vast improvement over piston-engine helicopters. The turbines weighed less and churned out more power.

The Bell UH-1's proper name was the Iroquois, but hardly anyone called it that. The nickname Huey came from the pronunciation of the chopper's original designation, the HU-1E, for "Helicopter Utility—1 Experimental."

In the years ahead, the Huey would become the force behind all airmobile operations in the war. Bell Helicopter would build more than 10,000 through 1975, and two-thirds were used in Vietnam. They could ferry troops, drop them off and pick them up in cramped spaces, resupply them, search for and rescue the trapped and wounded, fire guns, rockets and a grenade launcher. They could serve as an airborne platform for reconnaissance, for commanding and controlling movements in battle and for directing fire on the enemy. The crew consisted of a pilot or aircraft commander, a co-pilot, a crew chief and a door gunner.

Hueys had nicknames based on how, or whether, they were armed. One with two M60 door guns was called a slick because of its unadorned appearance and was used for transport. If it had two rocket packs of twenty-four rounds, it was a hog. If it had a nose fitted with a forty-millimeter automatic grenade launcher, it was a frog. If it was used to evacuate the wounded, it was a dust-off.

In fifty hours of flying time, during contact training and the first phase of tactical training called TAC 1, Nicky and the others in his class got familiar with the UH-1. They learned to take off and land safely, to sling heavy loads below them and to land without power in an emergency. One of the toughest challenges was learning the pre-flight checklist of more than a hundred items. In TAC 1, students learned formation flying, tactical maneuvers, how to deliver and pick up troops. Always in pairs, called stick buddies, they were graded together on how well they navigated and reached their checkpoints.

As stick buddies sharing the cockpit of a slick, Nicky and Tony alternated as pilot. In the checkpoint exercise, they did what they were supposed to do and watched for landmarks such as water towers, lakes, rivers and intersections. They entertained themselves at the same time with the ADF radio compass, an automatic direction finder with a frequency range that included AM radio stations.

"We'd tune in to a commercial station out of Montgomery and fly around listening to rock 'n' roll music," Tony said. "We'd know where we were going. We'd call in our checkpoints. We'd do that and fly around listening to the radio and have a good time."

TAC X was tactical exercise, another fifty hours in the air to refine what they already had learned, practice night formation flying, go on simulated combat operations. The idea was to acquaint the students with the kinds of situations they'd face in Vietnam, courtesy of instructors who'd been there. Tony remembered flying into remote areas, camping in fields, getting up at a moment's notice in the middle of the night, pulling his boots on and cranking up the aircraft. He'd fly to Fort Benning in Georgia to pick up troops and move them around, and to Eglin Air Force Base in Florida.

Knowing they were soon bound for the real thing in a war zone didn't overly concern them. "I don't think the longevity of aircraft pilots at that time was very high, but we never talked about that," Tony said. "Everybody knew that a lot of pilots didn't come back. But you have a nineteen- or twenty-year-old kid, he thinks he's invincible."

Skip recalled a fair amount of gallows humor about the prospects of getting back from Vietnam alive. "You had to kid each other," he said. "You knew each other better than brothers. You'd kid and razz, but you knew where to draw the line.... So all your fears, you had to have them out, and there was a dark humor about it. Everybody would kid about not coming back, but you couldn't dwell on it."

The further the student pilots advanced, the less stifling the military regimen they had to follow, and that was a welcome change. Unlike Forts Polk and Wolters, Fort Rucker was near several communities—Dothan, Daleville, Enterprise and Ozark. Enterprise was just outside the gate and was a favorite place for the fliers. They could hang out at the hamburger joints, just as they once did at home. "It was a nice, clean town," Tony said. "You could park your car and leave it unlocked and nobody would bother it."

Dothan, to the southeast, was the biggest town around, the home of a nightclub called the Oasis, where Tony went a few times. But Nicky, true to Terri, wouldn't go along even though Skip and Tony needled him. "We'd kid Nick about going out on the town," Skip said. "He never would. She had his heart and soul."

Just about every weekend they had the opportunity, Tony and Skip made the four-hour drive home in Skip's Chevelle, bringing Nicky along to stay with Tony's family. There in Rossville, tucked between the Civil War's Chickamauga battlefield and Chattanooga, the city once called the "Gateway to the Deep South," Tony's mother would serve Nicky what she called "real southern cooking"—green beans, fresh corn, cornbread and fried okra. Sometimes she'd "fix a beef." In the sultry heat they would drink iced tea. "I served him a banana pudding once and he said he'd never had such a thing and he liked it," Jewell Viall said.

Tony's youngest sister, Debbie, a teenager then and still living at home, became Nicky's friend. Sometimes the other guys would go out and Nicky would stay home with Debbie and her friends. "I guessed that he had wanted a break from all that male companionship," she said. "I do not recall any of our conversations, only that he smiled and flushed shyly in response to our teasing. I liked him very much."

On May 1, a month before the student fliers were to get their wings, the Department of the Army, Headquarters, United States Army Aviation School at Fort Rucker handed down Special Orders No. 115. Assuming the students would graduate, the commanders assigned Venditti, Viall and Vachon to the 16th Combat Aviation Group and ordered them to report June 28 for duty in Vietnam. To Skip's bitter disappointment, he would not be going to the war for which his pilot training had prepared him—and where his younger brother was fighting as a Marine—but to Korea to join the 38th Replacement Battalion. He would fly choppers for the Korean Military Advisory Group, providing what he called "a taxi service for colonels." He said he was one of only two in his class not assigned to Vietnam.

Given the chilling news out of the war zone that May, Skip was luckier than he realized. *The New York Times* revealed the secret B-52 bombing raids in Cambodia. The next day, May 10, troops from the 3rd Brigade of the 101st Airborne Division and supporting units began an assault on Ap Bia Mountain in the A Shau Valley, northwest of

Da Nang and near the Laotian border. Americans suffered 400 casualties in ten days of battle so grinding that it took the name Hamburger Hill.

By this time, Nicky was seeing the results yet again of what he could do when he applied himself intensely to a task, as he had done as a young baseball pitcher. Despite the flood of uncertainty in his letters home, he had focused on the job at hand and willed himself to succeed.

On June 1, he and 137 others in Class 69-13 were discharged from their enlisted status with the regular Army. The next day, they were appointed as warrant officers in the Army Reserve for an indefinite term. Though they were reservists, they were committed to active duty for three years as obligated volunteers. After that, they could quit or volunteer for indefinite status, meaning there was no set expiration of service and they could leave at any time.

Nicky cherished his new, privileged rank. It met his need for respect that had been so rudely disregarded on that high school map three years earlier.

On June 3, his stepmother proudly pinned a silver, wing-shaped metal insignia onto his Army-green jacket—the wings that identified him as a pilot. Aunt Bert had flown down from Malvern for the Fort Rucker graduation ceremony. She was the only member of his family to attend. Bert felt honored but bemused. "When he wanted me to go to Alabama," she recalled, "I said, 'Nicky, what about your mother?' He said, 'I don't want her to go.' I never asked him why."

Aunt Sally told me that she and her husband, John, a plumber, didn't go because they couldn't afford the trip. Uncle Louie didn't go simply because he didn't like to travel, though earlier in their marriage, he and Bert had gone to Florida on vacations. Now whenever the idea of traveling came up, Louie would resist, insisting, "Somebody's got to stay here and take care of the house!"

Aunt Bert gave me some pictures from graduation day. One shows Nicky, Tony and Skip mugging for the camera in their dress greens, or Class A uniform of a green jacket and tie. Nicky is in the middle, one hand on Skip's shoulder, the other on Tony's. Another photo has Skip with his dad and Tony with his dad and grandfather. Still another shows the graduates posing with their arms around their grinning moms, who wear dresses and—except for Bert—have their hair teased, the beehive style of the day. Bert's blond hair is up but not towering.

On June 8, five days after the ceremony, President Nixon, meeting

Edwin A. "Skip" Smith, Jr. (left), with his mother, Elsie Smith; Nicky Venditti (center) with his stepmother, Bert Venditti; and Jerald A. "Tony" Viall with his mother, Jewell Viall, on June 3, 1969, the day the men won their wings at Fort Rucker, Alabama.

with South Vietnam's President Nguyen Van Thieu on the Pacific island of Midway, announced the first withdrawal of American troops from Vietnam—25,000 soldiers. U.S. troop strength had peaked at 543,482 just five weeks earlier. Now Nixon was angling to turn the fighting over to the South Vietnamese. But in what Tony and many others saw as a weird joke, replacements kept going to Vietnam during the pullouts.

"It was political hogwash," Tony said, "peculiar, to be nice about it."

Fort Rucker would continue to turn out mostly helicopter pilots for the next three years. By the time the United States had fully withdrawn from the war, many thousands had passed through its gates and the overwhelming majority had gone to Vietnam. Of the 40,000 American helicopter pilots who served in the war, 2,202 were killed.

Among them, a dozen members of Class 69-13.

Chapter 6

Home and Unease, June 1969

For nine minutes, Nicky lives.

He seems older than his twenty years in the sure way he carries himself, smarter and tougher after a year of military regimen. His brown hair, curly like his dad's, is trimmed Army-short. The crisp khaki uniform fits snugly on his wiry five-foot-ten-inch, 158-pound build. Shiny, newly won silver wings identifying him as a pilot decorate his short-sleeve shirt.

He and Terri Pezick walk together across the lawn in front of his family's rancher outside Malvern. Terri, a slender eighteen-year-old with butterfly-frame glasses, wears a sleeveless, knee-length plaid dress with a bow at its rounded collar. Her light brown hair is pulled back in a ponytail and tied with a long white scarf.

Five inches taller than Terri, Nicky easily drapes his arm over her shoulder and pulls close the girl he plans to marry when he gets home from Vietnam. She puts her arm around his waist. In the house, he playfully pulls her onto his lap.

Three reels of silent, eight-millimeter color movie film contain the images shot by his dad, my Uncle Louie, in the days just before Nicky went overseas. After Louie died, Aunt Bert found them in a shoebox in the attic. Louie had known what was on them. "I'm never going to look at those," he once said when she asked him about them. No one did, and they were left unseen for almost thirty years, until I had the three-minute reels transferred to video.

Shot at different times during Nicky's leave, the movies present a snippet of suburban middle-class life in 1960s America. There's a picnic lunch of hot dogs, burgers and beer, with Louie and Bert, Nicky, his stepbrother Joe and best friend Charley Boehmler at the table. Louie, always the clown, makes faces and swivels his hips by the grill. Charley, a beefy ex–high school wrestler in a T-shirt, laughs while fending off Louie's attempts to lock him in a hold. Nicky lights a cigarette with his Zippo. He drives off with Terri in a pea-green '68 Camaro SS he'd bought in the South.

He is moving on my TV screen for a longer time than I remember seeing him when he was alive. I can't hear his voice, but I can observe his manner. As I watch him interact with friends and family, the ache of never having the opportunity to know him cuts deeper. I want to transcend time and space and step into this picture for a few precious moments—walk up to him in his parents' yard, look into his eyes and grasp his hand. "Hi Nicky, how are you?" At least that way, in June 1969, I would have connected with him before he was gone forever.

I feel angry, too, over the disturbing circumstances of his death just a few weeks later. Nicky would not be a cousin I'd know as an adult, someone I'd spend time with over the course of decades as we pursued careers and had families of our own. Something went terribly wrong at Chu Lai, and it robbed Nicky of a future and the rest of us of having him by our side. As I watch the film, it hurts to know this happy time unfolding in grainy images was Nicky's last. He had twenty-three days at home.

Occasionally during his leave he seemed to have a premonition of his fate, but not at the outset. Eager to see Terri after winning his wings at Fort Rucker, he stopped the Camaro only for food, gasoline and a few roadside naps on the 1,025-mile drive with Aunt Bert from southern Alabama to Malvern. Louie wasn't home when they walked in the back door. They found a greeting from him taped on the refrigerator door, a drawing he'd made of an airborne Army helicopter and an American flag with the note:

> Officer Nick, you're a good man.
> Welcome home, Nick & Bert!
> We sure did miss you....
> Bert, make us something to eat, clean the house
> and wash me some clean clothes.

Proud father that he was, Louie used the following days to fuss over his son. He took photos and the home movies and presented Nicky to his pals at the veterans' posts. "My dad wants me to put my uniform on and go down with him to the VFW," Nicky told his friend Charley Boehmler. He wasn't particularly comfortable doing that. "He felt a

Nicky as a newly minted Army warrant officer in June 1969 in the home of his father and stepmother outside Malvern. He was on leave just before departing for Vietnam. In a few weeks, soon after his arrival at the Americal Division base at Chu Lai, he would die at age twenty (courtesy the collection of Bertha and Louis Venditti).

little embarrassed that his dad wanted to parade him around in the Legion and VFW with his uniform on, but he was happy to make his dad proud," Charley said. "And he *was* somebody at that point. He had that officer tag."

Louie took pictures in their living room of Nicky in his dress blues—blue jacket with warrant officer bars on the shoulders, silver wings pinned on, a white dress shirt and bowtie. There was Nicky with his flight helmet on, visor down over his eyes; of him with his arm around Bert, and both of them grinning. Someone snapped a shot of Louie and Nicky, man to man, shaking hands.

One day, Nicky asked Bert's advice on choosing the beneficiaries of his $10,000 GI insurance, which the Army would pay if he were killed. "He stood by that stove, and I sat on that chair," Aunt Bert told

Nicky with his dad, Louie, at home in June 1969 while he was on three weeks' leave before reporting for duty in Vietnam. Louie, who had been a ground crewman with the 479th Fighter Group during World War II, was proud that his son had become an Army pilot (courtesy the collection of Bertha and Louis Venditti).

me in her kitchen. "He said, 'I've got this insurance, and I want to make sure that Terri's taken care of....' I said, 'Well, Nicky, you have a brother that might need some help. Why don't you share it with him and Terri, half and half?' He said, 'Good idea,' and that's all I heard, and that's the way it went."

Nicky used much of his time at home to visit relatives and friends across Malvern and other places in Chester County he had come to know in his twenty years. By and large, they were happy celebrations, with everyone glad to see him. He and Terri were invited to dinner at the East King Street home where my grandparents moved their family in the late 1930s, where Grandmom had fixed meals for Nicky and served him coffee the many times he stopped in. It was Aunt Patty and Uncle Jimmy's house now. Grandmom had died, and Grandpop was still living there but in poor health. Patty, the youngest of my grandparents' dozen children, and Jimmy, the Korean War–era veteran who had shown so much anger after Nicky's funeral, took care of Grandpop.

Following a macaroni dinner and pleasant chatter, everyone walked out onto the porch. Jimmy gave Nicky $10 and all said goodbye. Patty cried and told her husband, "We're never going to see him again." It was just a feeling she had.

Across town, another aunt conferred excitedly with Nicky on plans to mark his return from Vietnam the following year. "We were talking about when he got home, and he was going to get married, and I said we would have a thing in the back yard, and we'd have music," Aunt Dutchy said. "We'd put down wooden floors so people could dance. Oh, we had big plans!"

As the date of Nicky's departure neared, and his impending march to a savage war crowded his thoughts, he gradually sank into melancholy. He was not saying hello anymore; he was saying goodbye.

He stopped by the red-brick Malvern Public School to see Josiah Hibberd, the principal and teacher who had made grammar school fun and once had been like a second father to him. Nicky found him outside the old school at First and Warren avenues, loading grills and baseball gear into his Plymouth station wagon for a school picnic at Valley Forge. Years after taking Nicky on dozens of similar trips, Hibberd was still at it. The thought must have filled Nicky with admiration for his old mentor.

Eager to share his concerns with someone he trusted and whose opinion he valued, Nicky let on that he was leery of going to the war. Hibberd, a World War II veteran like many of his generation, detected the edge in Nicky's voice. The teacher understood the young soldier's anxiety about the dangers that lay ahead and sought to reassure him. He emphasized the need to stay out of trouble and on the alert. "I told him, 'Keep your nose clean, your ears pricked and your eyes open, and you'll be all right.' He just smiled."

For reasons unexplained, Nicky didn't call or visit Mary Anne Wallace, whose friendship he had nurtured and valued. A meeting with her might have been awkward, considering that he and Terri were moving toward marriage and that Terri was suspicious of Mary Anne's motives.

Nicky had written to Mary Anne every week that he was in training and sometimes called her. She responded to every letter she got from him, and sent him packages of chocolate chip cookies she had baked. And though their letters were often just one page and unremarkable, mere chitchat, the contact worked for both of them. It was easier to keep their friendship going when Nicky was hundreds of miles away than when he was at home, with Terri around.

The last time Mary Anne saw Nicky was at Christmastime 1968, while he was home on leave. He came to her house and they chatted, but not about his prospects in Vietnam. He understood that she was uncomfortable talking about that, and never forgot that she had opposed his signing up.

Perhaps in not contacting Mary Anne during his last leave, he wanted to spare her emotions as well as his own. Many years later, when I told Mary Anne that Nicky had been home for a few weeks that June—something she hadn't known—hurt flashed across her face. She couldn't fathom why Nicky had passed her by.

He didn't face her, but he hadn't put her out of his thoughts. When he left for overseas, he took along a St. Christopher medal Mary Anne had given him a year earlier, before he left for boot camp. She had prayed that the patron saint of travelers would watch over him wherever he went. Given his fears as Vietnam loomed, he was not about to leave behind this symbol of God's loving care.

A further sign that Nicky was sorely conscious of his mortality, and that Terri was uppermost in his heart, came when he visited her best friend just before his twenty-three days were up. It was more than a farewell.

Bobbie Stiteler was one of Terri's classmates at Great Valley High and knew Nicky well. He used to pick up the two girls after school for rides home. Bobbie and Terri were confidants, and Bobbie's fiancé was one of Nicky's street-racing buddies. Nicky surprised Bobbie, showing up at her home.

"I won't be coming back," he told her.

"Oh, you're just scared," she answered. "Everyone feels that way before they go to Vietnam."

"No, I know I won't be coming back," he insisted. "Think whatever you want, but I won't be back."

His tone troubled her. She didn't know what to say.

"Can you do me a big favor?" Nicky asked. "I want you to be there for Terri, because she'll need you."

Bobbie assured him that she would. He hugged her and said, "You've been a good friend to me." Then he went away, leaving Bobbie deeply upset.

The night before Nicky left Malvern, he and Terri went to a drive-in with Charley Boehmler and his wife-to-be to see the World War II action adventure *Where Eagles Dare*. In the film, Allied commandos go on a mission to rescue an important American officer from the Nazis. "It was a real shoot-'em-up and a mistake for us to see," Charley said. It left Terri in tears.

When they said goodbye, Nicky confided to Charley: "I may not come back. Take care of yourself."

Charley was reassuring.

"Oh Nicky, you're lucky, you're always lucky! You're going to be the best man at my wedding. You've got to come back."

On the afternoon of the next day, Charley and Aunt Bert took Nicky to Philadelphia International Airport. His twenty-three days were up. As they drove past Sally's house, Bert crouched down in the back seat. "Don't let my mom see you," Nicky warned. He hadn't invited her to go along.

He wouldn't let Terri go, either. He wanted them to part in a warm and familiar place, his parents' home, not in the cold environment of a big-city airport. Terri stayed behind and napped on Nicky's bed. She cried when they said goodbye. She saw he had opened a Bible and underlined "Psalm 23" in red, but didn't think anything of it. It was only after Nicky's death, when she spoke with her friend Bobbie Stiteler, that she thought Nicky might have foreseen the end.

Bert felt the same: "I think he knew."

At the airport, she cried as her stepson boarded a plane to Seattle.

From the Pacific Northwest, the next leg of Nicky's trek across the world began at McChord Air Force Base, just south of Tacoma, Washington. About 5 a.m. on July 3, he got on a chartered commercial flight destined for Vietnam by way of Alaska, Japan and the Philippines. He had seen his Army buddy Tony Viall briefly in Seattle, but Tony would get on a later flight for the long journey across the Pacific.

During the stopover in Japan, Nicky sent his mother and stepfather a color postcard of a geisha girl. "We have an hour stop here before we leave for 'Nam," he wrote on the back.

On the Fourth of July, he arrived in South Vietnam. The next day, he wrote to his dad and Bert:

Well I arrived at this wonderful place called Vietnam yesterday at 3.... It was about
100 degrees. I still can't believe I'm here, but when I look around I get more
assured I am! ... I haven't seen Viall since I left Seattle. But he should get here
before I leave. Oh I'm at Cam Ranh Bay replacement center right now. It's about
150 miles from Saigon. It's probably the safest place in Vietnam. Too bad I can't
get stationed here.... See you in 363 days.

Cam Ranh Bay was a huge U.S. base on the lower end of South Vietnam's
bulge into the South China Sea, down from the seaside resort town of
Nha Trang. To the south, the land curved westward to Saigon. In 1965,
Americans turned the bay's harbor into a deepwater port with cargo
cranes, prefabricated concrete piers, warehouses, fuel storage tanks,
hospitals, barracks, and a runway 10,000 feet long.

One of Nicky's Fort Rucker classmates also arrived at Cam Ranh
Bay on the Fourth of July—Thomas Sled. A native of Chicago's South
Side who grew up in Whittier, California, Tom was a helicopter pilot
by a timely stroke of good fortune. Though he had graduated with Tony
Viall, Billy Vachon and Nicky, his Army experience before flight school
was different from theirs. He had taken a less direct route to the rank
of warrant officer by not learning to fly immediately after basic training,
as the others had done and as he had wanted to do.

He had been a rudderless college student, getting poor grades at
the University of San Diego because he liked to party. One weekend in
San Diego's Mission Bay, he bought a helicopter ride for fun, an experience that led him to quit school and join the Army to fly choppers.
As an aviator, he would be following in his father's footsteps. His dad
was a B-17 navigator during World War II.

At Fort Ord, California, Tom went through two months of basic
training, then four months of advanced infantry training. Along the
way he took and passed the tests to get into flight school but never
heard anything about it. On the day he was being processed to ship
out to Vietnam as a "ground-pounding" armorer, an infantryman who
maintains and repairs small arms, he got word that he had been
accepted into flight school. He went through Forts Wolters and Rucker
and got his wings in June with Billy, Tony and Nicky in Class 69-13.
Like the others, he was assigned to the 16th Combat Aviation Group,
which was attached to the Americal Division.

Tom's Trans World Airlines flight touched down on Cam Ranh
Bay's sandbar at sunrise, eight hours earlier than Nicky's plane. When
the jetliner taxied to a stop, he got a rude greeting. "I was sitting right
by the door, so when they opened it, a returning sergeant sitting next

to the door let me be the first to get off the plane." The temperature outside was 102, the humidity 98 percent. Tom paused. "What the hell is that smell?" he asked. The sergeant behind him said, "Shit, sir. We burn our shit here, sir."

Burning human waste to dispose of it was the standard practice at base camps and fire support bases across the country. Fifty-five-gallon drums were cut in half and placed in latrines, similar to the outhouses in use before indoor plumbing. Every day, a soldier was assigned to pull out the drums, pour kerosene into them and set the mixture afire. To get a thorough burn, he had to stir the contents with a section of pipe or a sturdy piece of wood. The result was plumes of black smoke and a foul odor.

At the replacement center, Tom processed in through the paperwork. "We had a thirty-second physical, saw several fifteen-minute and thirty-minute movies about VD, how people can be traitors, on cleaning your M16, on taking malaria pills, and tooth care. After that, we sat around waiting for some of the other guys to get in before we would be transported up north. We spent a lot of time drinking beer in the officers' cantina, a building the size of a two-car garage, with one door and no windows, but it did have eight air conditioners built into the wall, a jukebox, a bar and twenty slot machines."

On July 5, Tony Viall arrived at Cam Ranh Bay, also on a commercial flight. As soon as he stepped off the plane into the roasting air, he knew he'd come to the war. He boarded a bus that had wire screens over the windows, a means of preventing the enemy from tossing explosives inside. Across the bay, he saw helicopter gunships circling over a target of some kind, blasting away. He found Nicky by chance at the replacement center, where soldiers' orders were checked and transportation to their assigned areas arranged. The reunion took place soon after Nicky mailed his letter to his dad and Bert. The two friends had a few beers together.

Hours later, in the early morning darkness of July 6, they were rousted from sleep, got their gear and took a bus to the flight line on the runway with Tom and others. They waited to clamber into the gut of a burly C-130 Hercules transport.

"After drinking heavily and getting over the initial shock of being in-country, we were so far having a good time," Tom said. "But then while we stood there, they unloaded thirty-seven body bags off this C-130 that just came in. That sobered us up and brought us back to reality."

The C-130 has a door in the rear that lowers and serves as a drive-up ramp for loading and unloading. Crewmen moved a huge earthmover tire up the ramp and placed it in the middle of the drop door. They also loaded a few pallets of helicopter rockets. The men got in the back, and the ramp went up. Tony and Nicky sat together on the tire, while Tom made himself cozy in the center of it to sleep.

The thrumming roar of the cargo plane's four turboprop engines kept Tony and Nicky from talking. The journey up the coast would take an hour and twenty minutes. They were heading for a vast American base along the South China Sea in the northern part of South Vietnam.

They were heading for Chu Lai.

Chapter 7

"Isn't worth a nickel,"
July 1969

Tom Sled and some buddies were relaxing and enjoying relief from the heat and the odor of the base by swigging beer at the Americal Division officers club on a hill of the Ky Ha Peninsula. They looked down on tranquil Dung Quat Bay and noticed a tiny boat navigated by a Vietnamese man seemingly struggling with his craft. The boat was too close, a breach of security, but hardly looked like a threat.

Suddenly a mortar exploded on the beach below the club. The man had fired it but had punched a hole in his boat with the launcher and was sinking fast. In a minute, a Huey gunship from the Ky Ha Helipad lifted off, and its crew spotted the man trying to swim away. The gun-laden hog fired a torrent of rounds, and in seconds all that remained was spray.

This was everyday life at Chu Lai, headquarters of the largest Army unit in Vietnam. The Americal Division had 20,000 troops in six major combat commands, including three infantry brigades. Officially the 23rd Infantry Division, it was created during World War II from a task force sent to defend an island east of Australia called New Caledonia, a former French penal colony. The name (pronounced Ameri-CAL) fused "America" and "Caledonia."

New arrivals faced far more than the incongruous interplay of lethal gunfire and beer. They had to adjust to the oppressive heat and humidity that had enveloped them since they landed at Cam Ranh Bay. They had to get used to the thunder of the Marine Corps' jet bombers—F-4 Phantoms, A-4 Skyhawks and A-6 Intruders—taking off and landing at all hours. They had to bear the stench of human waste burning in fifty-five-gallon drums. Another daily reality was the nearness of the enemy, who fired mortars and rockets onto the base from just a few miles away. Some even were in the GIs' midst, infiltrators among the Vietnamese civilians who did laundry, cut hair and cleaned hooches.

The month before Nicky's arrival, a cadre of the North Vietnamese Army fired a cluster of Soviet-made 122 mm rockets from beyond the Chu Lai perimeter. One of them hit the Vietnamese ward of the 312th Evacuation Hospital, on the Ky Ha Peninsula just south of division headquarters, killing a Vietnamese child and an Army nurse, First Lieutenant Sharon Lane, from Canton, Ohio.

Chu Lai had been an enemy target ever since Americans created it in the mid–1960s, when Navy Seabees built an airstrip on orders from a Marine general, Victor "Brute" Krulak. Chu Lai wasn't a Vietnamese term, but the pronunciation of the Chinese characters for "Krulak." The war's first major battle between Americans and the Viet Cong took place in the summer of 1965, when the guerrillas tried in vain to seize the airstrip. Two years later, the Americal Division came to Chu Lai to ease the pressure on the Marines fighting near the Demilitarized Zone, the 17th Parallel dividing North and South Vietnam. The base was part of I Corps, the northernmost tactical zone in South Vietnam. After 1966, more than half of all Americans killed in action died there.

That zone included My Lai, where an Americal unit slaughtered hundreds of villagers on March 16, 1968, a year and a half before Nicky landed at Cam Ranh Bay. Commenting on why My Lai happened, Vietnam veteran B.G. Burkett and Glenna Whitley criticized the Army in their book *Stolen Valor*. They said commanders of other units used the Americal Division as a dumping ground for misfits and troublemakers. "Among all the major units in Vietnam," they wrote, "it had the reputation of being the most disorganized, with some companies populated by less-than-sterling soldiers."

Tony Viall had heard otherwise. Soldiers he'd talked with had spoken highly of the division, giving him the impression it had an admirable fighting spirit. Its troops paid heavily for their role in Vietnam. Ultimately, nearly 4,000 died.

When Tony and Nicky arrived at Chu Lai to join the 16th Combat Aviation Group, they reported to the personnel office at division headquarters on the peninsula. With Billy Vachon and Tom Sled, they asked to be assigned to the 176th Assault Helicopter Company, an outfit they'd heard good things about—most important, that its members were proud and professional and got the job done. Three months earlier, six of its crews had received twenty-one medals for a daring rescue near

Thien Phuoc. No other Americal unit had gotten more honors for a single combat action. The men of the 176th linked the unit to the spirit of the American Revolution, calling themselves the Minutemen. Its gunships—one of the company's three platoons—were the Muskets.

Though Warrant Officers Sled, Vachon, Viall and Venditti wouldn't get their assignments until after orientation, they had high hopes of joining the 176th. For now, they settled in for a week at the Americal Combat Center, just down the coastline from the evacuation hospital where Sharon Lane had been killed. Training and temporary housing at the center were provided for the FNGs, the fucking new guys.

The center stood along a beach of bright white sand about four miles south of division headquarters. A sign on a fence assured the trainees they could have "Power Thru Knowledge." An asphalt road ran along the shoreline less than fifty yards from the water. Though concertina wire was stretched along the road's seaside edge, it wasn't much of a deterrent, mostly just a single strand that almost anyone could breach.

New troops stayed on the beach in barracks with bunks and bare mattresses that were sweat-stained and smelly. There were several classrooms, an aid station, an amphitheater, a motor pool of trucks and a "gas chamber" that trainees had to run through. The buildings, single-story plywood structures with corrugated tin roofs, dotted acres of beach and low sand dunes. Sand was everywhere. Empty rocket pods from helicopter gunships served as piss tubes, or outdoor urinals.

Metal bunkers similar to large culverts offered protection from incoming rockets or mortars. The bottom half was buried in the sand, the top padded with green sandbags. Rocket and mortar attacks were considered an irritation rather than life-threatening; they were infrequent, and the thirty-six-square-mile Chu Lai base was a broad target. Though there were casualties at times, the odds of being killed or wounded were long. Still, new and inexperienced troops got the jitters when they heard explosions.

Longer-term residents left their nerves behind for what recreation they could grab. Pennsylvanian Frank "Max" McLaughlin, a mess hall baker while stationed there, was a lifeguard at the center's beach and liked to surf. One day he got into a contest with two Californians as they challenged one another on high surf kicked up by a storm at sea. Max's board dipped in front and he was catapulted over it. The board came down on the back of his neck, knocking him unconscious briefly and pushing him down into the water. "I remember thinking that I was in Vietnam and might have been killed while surfing." He floated to the surface and bodysurfed to the beach.

During their week of getting over jet lag and adapting to the tropical climate, new arrivals heard instructors go over the basics, such as the importance of good hygiene and keeping weapons clean. They walked a trail rigged with dummy mines, which were set off to give them an idea of what to look out for. They learned about the enemy's latest tricks and to always expect the unexpected.

At the gas chamber, the idea was to give new troops a dose of what they'd experience if chemicals were used against them. An instructor would take CS powder out of a bag, put it in a metal pan and light it. The smoking gas would fill the rectangular room, which had doors at the far ends. Soldiers had to enter the building at one end, without masks, and find their way to the exit at the other end. It took them just a few seconds to rush through the building, but that was long enough to get a good whiff of the hot, sticky gas. By the time they reached fresh air, their eyes and skin were burning.

On Sunday, July 6, Nicky wrote to his dad:

> I'm sitting at the Combat Center here at Chu Lai. I'll be here about six days before I'm shipped out to my unit.... There are choppers and jets flying all over the place here. I'm sorry this is a little sloppy, Dad, but it's hotter than hell here. It makes Fort Polk seem air-conditioned.
>
> Well I'll let you in on the situation up here, Dad. It's not too good.... The lieutenant who briefed us said they expect an offensive, but do not know when....
>
> That's all I can let you know for now. Besides, I wouldn't tell you anymore anyway, because you'll worry your head off.
>
> How are my women and my car doing? You know you have to take care of both of them till I get home. If Terri needs anything, get it for her, OK.... Take care, Dad, and don't worry about me.

Nicky did give his father something to worry about in a letter the next day. About three in the morning, while sleeping in their barracks on the beach, Nicky and Tony had their first brush with danger.

> Mortars started coming in. I heard the first two rounds hit and saw everyone run like hell. So I rolled over in bed, and after awhile, the alert siren blew. I decided I'd better find a bunker. You would've laughed if you saw Viall. He jumped out of bed, fell out the door and low-crawled to the bunker. That was the fastest I ever saw Viall move.... See you in 361 days (I think).

That same day in a letter to his mother, Nicky left no doubt about his feelings: "Mom, this place is lousy. I can't even see why we are here because Vietnam isn't worth a nickel."

But he was stuck there for a year, and sometimes the absurdity of this new and temporary home, where leisure and violence played off each other, left him laughing or bewildered or both. It was as if he had nodded off into some quirky dream. That's what it was like one day when he and Tony were kicking back at the officers club and heard explosions and gunfire from down the hill. "We thought: What a crazy war this is," Tony said. "We're sitting here drinking beer and people are shooting at each other."

Another surreal scene had them dashing every which way in the midst of a mortar barrage. They were walking on the beach in daylight and out in the open, which they'd been warned not to do, when "Charlie" decided he was going to throw some stuff in, as Tony put it. "I turned and took off in one direction. Nick took the other. We thought, this doesn't work, and we turned around and went right past each other. We did this a couple of times, like Keystone Kops. Then we said, hell with it, and walked on."

When they got into mischief that others seemed to completely overlook, as if perpetrated by ghosts, it added to the atmosphere of unreality. Tony shoved a lieutenant as they walked on the narrow steel panels laid down to cover mud and wet sand when it rained. The next day, the man had his arm in a sling. One morning as they lay in their bunks, a lieutenant came in to wake them up and Nicky—startled or half-drunk—punched him hard in the face. "I was thinking," Tony said, "I gave a lieutenant a broken arm, and now Nick's given one a broken nose." Yet nothing came of either encounter.

On Thursday, July 10, the fourth day of their orientation, they ate an early breakfast in the Combat Center's mess hall. On their schedule was a morning session on grenades at a firing range off the base, at Landing Zone Bayonet. A severe storm warning was in effect for Chu Lai until 6 p.m. It was Tropical Storm Tess, churning 300 miles out in the South China Sea and threatening winds up to sixty-three miles an hour, thunderstorms and ocean breakers to eight feet. Across Vietnam, there had been a three-week lull in combat. The day before, the only news from the war zone was that enemy ground fire had downed two Hueys near Saigon and Da Nang, killing five Americans. Elsewhere, a battalion of the 9th Infantry Division had redeployed to the States, the first troops to withdraw under President Nixon's plan to have the South Vietnamese take over the fighting.

Americal troops' contact with the enemy on July 10 was "light and scattered," according to the journal kept at the division's Tactical Operations Center. Two soldiers were killed and fifteen wounded in fighting that also left twenty-two Viet Cong and two North Vietnamese Army soldiers dead, the journal writer noted.

Despite the storm threat, the sun shone hot and bright as Nicky, Tony, Billy, Tom and other replacements piled into the open beds of two-and-a-half-ton trucks—deuce-and-a-halfs, the men called them—and rode away from the Combat Center, westward across the sandy base to Highway 1, the French-built road that formed the inland edge of the camp. Two miles south of Chu Lai's main gate was the entrance to LZ Bayonet. The trip to the landing zone took less than half an hour.

The firing range and orientation building lay on the western edge of Bayonet. Low plywood buildings and piles of sandbags revealed the presence of troops who were settled in for the long term. Among the buildings were the brigade headquarters and Tactical Operations Center, which sprouted a cluster of radio antennas. There were mess halls,

The orientation building (right, background) at LZ Bayonet. The soldier on the left is Don Stuhr of the 1st Battalion, 14th Artillery, who sent me the photograph and believed it was taken in fall 1969. He didn't remember the name of the other soldier (courtesy Don Stuhr Collection).

hooches, "clubs" that offered beer and music, a barbershop and a PX, or post exchange. Observation posts lined the hills.

The orientation building was twenty-five feet wide and twice as long, with screen doors at each end and a corrugated tin roof. Behind it lay the firing range, nearly flat ground stretching 200 yards to a ridge of high hills that formed a natural backstop. Knee-high scrub brush blotted the hillside. Guys went there to practice firing M16 rifles, M60 machine guns, M79 grenade launchers.

"We were in a convoy of several trucks," Tony said. "I remember riding, and there was probably conversation about what we were going to be doing. When we got there, there were all these guys getting off the trucks and being put into formation, probably a very loose one. The four of us just wandered into the building where this was going to take place. We went in the door and sat down up front.

"I don't know why we chose that particular table."

Chapter 8

"We gotta get out!"
July 10, 1969

Let's get this over with, Tony Viall thought.

Bored and drowsy from the humid heat, he felt moderate annoyance as the Army instructor greeted the class and started to talk about grenades.

Grenades? That's for the ground-pounders, Tony thought, the guys who guard a perimeter, go out on an ambush or file along a jungle path and get into a firefight. That meant many of the several dozen troops seated behind him, not fliers like him.

No, the pilots belonged with their helicopters, not in this class geared to the infantry. The job Tony, Nicky Venditti, Billy Vachon and Tom Sled were in Vietnam to do wasn't on the ground; it was from above. Just let us do it, Tony thought.

Minutes earlier, when the four of them arrived at the orientation building, they had hopped off the trucks and ambled inside ahead of the other soldiers. It was about 10 a.m. They picked a bench in the front row and sat together, their backs to a wooden table—Tony on one end, Nicky beside him, Billy beside Nicky, Tom on the other end. The other guys filed in and settled at the remaining tables. They faced a foot-high stage with a lectern and a table, on top of which lay a box containing the instructor's grenade and other props.

Everyone wore a "steel pot" helmet, fatigues and boots. Tony took off his helmet, laid it down so he wouldn't have to hold it. Others removed their helmets or pushed them back from their foreheads.

The instructor was a sergeant in his early twenties, barely older than the guys in his class. Tony paid scant attention to him as his mind drifted from the dulling effects of boredom and the hot, heavy air. The minutes dragged by. He lolled in his seat, the words from the front of the room reaching him, drifting past, not registering.

Tony's torpor was not broken when the sergeant held up a fragmentation grenade. With his thumb pressed over the safety lever, he pulled out the safety pin ring but kept the grenade in his grip as he talked.

Here we go again, Tony thought.

He and his friends were used to seeing dummy explosives. In a scare class the day before, they walked a trail to get acquainted with the enemy's latest tricks with mines and booby traps. They tripped wires attached to fake explosives and saw how they were rigged.

Tony continued to be unimpressed when, about 10:15 a.m., the sergeant let go of the grenade. He tossed it as if fumbling it, but with some force. Tony noted the sweep of his arm.

"Things happen," the sergeant said.

The grenade's spring-loaded safety lever sprang off.

Tony of course knew the instructor wouldn't pull such a stunt inside with a live grenade. Others weren't fazed, either, as the grenade thunked onto the gray, asphalt-tile floor and rolled toward the tables.

One second...

From the other end of the pilots' bench, Tom Sled watched the grenade bounce off the wall to his right. It slipped under the table next to the one where the four warrant officers sat. One of the guys over there might have kicked it, propelling it onward.

Two seconds...

The sergeant looked out over the room and said something, but Tony didn't take note of it. He still wasn't even halfway interested. It might have been, "Is this the way you guys react to danger?" But Tony and the guys around him stayed put. They knew it was a joke. They weren't going to play the panic game. They wouldn't show any fear over this dumb little lesson.

Three seconds...

The grenade skittered across the floor, heading directly for the four friends on the front bench. It rolled between Tony and Nicky, going under the table at their backs. It looked as though it would end up by Tom's foot.

Tom reached down and behind the backless bench to pick up the grenade, figuring he'd toss it back to the instructor. But it arced away, curving across the floor.

It was still under the table when it stopped.

Four seconds...

Tom gave up on grabbing it, now beyond his reach. If someone were going to pick it up, it would have to be one of the men behind him, guys the pilots didn't know. Tom started turning to face the front. Beside him, Billy, Nicky and Tony hadn't budged.

Five seconds...

A thunderous *CRACK* sounded as a blinding torrent of hot jagged metal in a storm of smoke filled the room. The time-delay fuse had taken five seconds to burn down and set off a high-explosive charge surrounded by notched and coiled metal strips. When the grenade burst, the metal broke into a thousand quarter-inch bits that flew at terrific, deadly speed amid chunks of iron casing.

Tom went airborne, hurtling five feet forward and crumpling at the foot of the instructor's stage. Around him, guys stampeded to the doors, yelling and screaming. Dazed, his ears ringing and throbbing from the concussion, Tom grabbed someone and yelled, "We have to get to cover 'cause we have incoming!"

But his bloodied right leg failed him. Fragments had gashed it from mid-calf up to the thigh, severing a nerve at the knee. A piece of black metal stuck out of his right elbow. His left ear, which had been turned away from the explosion, was nicked and bleeding.

Someone seized his arm and pulled him, shrieking, "We gotta get out, we gotta get out!" Apparently, they thought, the enemy had zeroed in on them with mortars or rockets.

Tony tried to get up from the bench but slumped to the floor. Nicky lay conscious beside him.

Nick has the same problem I have, Tony thought. It's our legs!

Everyone else seemed to have disappeared. Tony couldn't see Tom or Billy.

Tony and Nicky pulled themselves along the floor, using their arms and dragging their useless legs. They left a trail of blood as they worked their way to the closest door ten feet away. Deafened, they didn't try to speak, but inched toward the door. When they got there, hands from outside reached in and grabbed them, first pulling out Nicky, then Tony. Blood spilling from their wounds smeared the hot dirt.

Soldiers from the surrounding base rushed to help the crowd of cut-up men. "What happened?" they yelled. Medics sped to the building in a jeep ambulance.

Tony watched one medic cut off his boot and wrap a tourniquet around his lower left leg, slashed below the knee. The thought shot through him: This is serious, I'm going to lose my leg! He realized, too, that hot metal had pierced his right foot. When he complained about the pain, a medic pulled off the boot, saw the blood and applied a tourniquet there as well.

His hearing began to return.

Nicky propped himself up on his right arm. Tony turned to see him. Their eyes locked. The look on Nicky's face was something Tony had never seen before, something he couldn't fathom. He was sure Nicky was coherent and recognized him, yet his expression seemed oddly vacant.

Someone pulled Tony back, blocking his view, but he heard a medic tending to Nicky say, "He's going into shock."

A Huey touched down. Soldiers lifted Nicky into a side cargo door. Its turbine engine pounding, the rescue chopper rose on roiling dust and nosed east toward Chu Lai. The base, only minutes away, just across Highway 1, had two fully equipped hospitals.

When a second chopper settled to the ground, Tony, Tom and Billy were placed on the deck behind the cockpit. Tony and Tom lay flat on their backs. Billy sat up. He seemed to be alert and was looking around, even though he had a puncture wound between the eyes. Tony didn't see a lot of blood. He couldn't tell whether Billy had any other serious injuries.

Looks like he's going to be all right, Tony thought.

Nicky was in the worst shape, with both legs hit and one really bad. But, Tony thought, his buddy for the last year, the all-around good guy he had trained with in boot camp and two helicopter schools, with whom he'd downed copious amounts of beer and shared countless late-night pizzas, would be OK. Leg wounds were survivable and besides, Nicky was confident, proud, athletic, the kind of man you look up to. He wasn't going to die.

The chopper with Nicky aboard vanished. Tony would never see his friend again.

Chapter 9

Bitter Pills, June 1969

Lynn O'Malley came to Chu Lai in the spring of 1969 after Army nurse Sharon Lane was killed in a rocket attack, the only American servicewoman to die from hostile fire in Vietnam. All at the 312th Evacuation Hospital were in deep grief for their fallen colleague, making the first few weeks of the terrible eleven months Lynn spent there especially difficult. From the start, the alien reality of Chu Lai unsettled her. She had just turned twenty-two and suddenly found herself thrust into a war zone, in a forbidding climate of the cruelest heat, in the filth and drabness of a cluttered military outpost. Always, there was the danger posed by the enemy with his rockets and mortars.

She had arrived in South Vietnam three days earlier, at Bien Hoa, site of a major base sixteen miles from Saigon. Now she was on Chu Lai's Ky Ha Peninsula, near a bluff overlooking the South China Sea, struggling to do her job in a hospital where she was a stranger and—with only five months of experience as a registered nurse behind her—uncertain of her ability. The people she worked with were still trying to fend off the shock of Sharon's violent death. At some level, Lynn understood their preoccupation with what had happened to one of their own, right in their midst, on a day like any other day in Vietnam, but that didn't make coping in this harsh new place any easier.

It was one more bitter pill for Lynn, who had felt tricked into coming to Vietnam. She had grown up in little Walnutport, Pennsylvania, and gone to nursing school at St. Luke's Hospital in New York City. Inspired by an aunt who had been an Army nurse for two decades, she signed up for the Army Student Nurse Program in the fall of 1967 during her senior year. The Army paid her $118 a month until she graduated the following May, after which she remained in the city, working at St. Luke's. That summer she passed the state board exams and became a registered nurse, making her eligible for the Army Nurse Corps. She was commissioned as an officer and, in the fall, took basic training at Fort Sam Houston, Texas. Then it was on to Fitzsimons Army Hospital in Denver, where she worked briefly in a medical intensive care unit. She had been assured she wouldn't go to Vietnam unless she volunteered, but that turned out to be untrue—she got orders for deployment in February 1969. It grated on her. She didn't want to be in the war.

She had landed at Bien Hoa on June 8, about a half-hour after Sharon bled to death when shrapnel from an enemy rocket sliced her carotid artery as she worked in the hospital's Vietnamese ward. The attack happened at sunrise. Sharon, twenty-five years old and in Vietnam just six weeks, died instantly. Lynn heard none of this from the chief nurse at Bien Hoa's replacement center, just that a nurse had been killed at Chu Lai. She got her assignment: the 312th Evac.

It would, in just a month, connect her to my cousin Nicky.

Lynn's last leg to Chu Lai did not go smoothly. Twenty-two sleepless hours of air travel from Travis Air Force Base in California, with stops in Alaska and Japan, had left her exhausted. At Bien Hoa's replacement center, she dozed for twenty-eight hours in an air-conditioned trailer and then met with the chief nurse. Now, two nights later, she stood with a large group of soldiers on a makeshift bridge over a ditch in Bien Hoa, waiting for a C-130 to Da Nang. Someone told her there was less chance that the aircraft would be fired on in the darkness. As they waited, one of the guys moved too close to her. She hadn't gotten used to close contact with others, even after three years of riding subways in Manhattan. She backed away from the soldier, lost her footing and fell into the ditch four feet below. She wasn't hurt and GIs rushed to help her. She was coated with dirt.

In Da Nang, she spent all night in the airport and left in the morning on a helicopter for the forty-five minute ride south to Chu Lai, where she reported to the 312th Evac. Close by the hospital, the bluff fell off sharply to a beach of bright white sand. Beyond it lay a turquoise sea that looked as soft as Vietnamese silk. And though it was nice to watch the sun rise over the ocean, the compound itself was an ugly mess of dirt and bleak Army buildings.

The 312th Evac* was an Army Reserve unit from Winston-Salem, North Carolina. The hospital consisted of Quonset huts with the latest

---

*The 312th Evac was being replaced by the regular Army's 91st Evacuation Hospital, which had been at Tuy Hoa in the Central Highlands. The transfer was completed Aug. 1, 1969.

**The 312th Evacuation Hospital on the Ky Ha Peninsula at Chu Lai in 1969.
The 312th Evac was an Army Reserve unit from Winston-Salem, North Carolina, and would be replaced by a regular Army unit, the 91st Evacuation Hospital (courtesy Lynn Bedics Collection).**

technology and could hold 150 patients at a time. They were American, South Vietnamese and enemy soldiers, and Vietnamese women and children. Most had wounds from gunfire, shrapnel or fragments. Others were sick. Malaria was the most common ailment.

Lynn met the chief nurse and assistant chief nurse, two women about her parents' age who jumped out of their seats and greeted her with hugs. They told her the hospital had been hit and a nurse and a patient had died, but nothing more about the tragedy. The assistant chief took Lynn to one of the two bachelor officer quarters, where she had a windowless room that seemed like a jail cell. It was on the first floor, on the side facing the hooches where the doctors lived and where *mama sans* hand-washed the staff's fatigues and bed sheets and hung them on lines to dry. As the weeks stretched into months, Lynn immersed herself in work in the intensive care unit and the recovery room.

The ICU was a Quonset hut with about twenty beds connected by an enclosed hallway to another Quonset hut, the recovery room, where patients stayed for a few hours after surgery. Both huts were air-conditioned. In the ICU, the worst-off Americans were tended to until they were stable enough for transfer to Da Nang. Some died before they could be moved; no one stayed more than a couple of weeks. Nurses and corpsmen assigned to the ICU also worked in the recovery room, and staffing was fluid. When Lynn was the charge nurse in the recovery room and the nurses in the ICU were busy, she assisted them. She got generous help from the other nurses, who taught her how to manage critically ill surgical patients—care that was outside her experience.

Thirty years later, sitting across from me in her office at a Veterans Affairs outpatient clinic, Lynn O'Malley Bedics' dark eyes pooled at the memories, and the ready smile with which she had greeted me faded from her smooth-skinned oval face. As her thoughts turned to

Second Lieutenant Lynn O'Malley of the Army Nurse Corps at a shelf of intravenous bottles at the 312th Evacuation Hospital, Chu Lai. Lynn was twenty-two years old when she arrived in Vietnam in June 1969. The next month, she tended to Nicky in the intensive care unit as he lay dying (courtesy Lynn Bedics Collection).

Chu Lai, her concentration on the past was so total that even though she looked directly into my eyes, it was as if I weren't there.

"We'd get patients with 100 percent body burns. They'd be conscious and live for days," Lynn said. The nurses gave them pain medication, intravenous fluids and emotional comfort, but these patients could not recover. They were kept in a separate area of the twenty-bed recovery room, which was less crowded than the ICU. And while these most horrible cases are uppermost in Lynn's memory, other patients had less severe burns and survived. She noted that most burn patients were Vietnamese and that all of the burn victims she recalled had been scorched by napalm, a flammable liquid used by U.S. forces.

There were soldiers with other grievous injuries. "I had one patient who lost all four extremities, and he died while he was there," Lynn said. A GI might be paralyzed from the neck down or horribly disfigured because of facial injuries. Every day, Lynn wondered what would happen to men such as these. How would they get by, if and when they made it home? If they survived, would they even want to live?

"It was really awful," she shook her head as if trying to banish the memory, and her voice rose in anguish. "I'm *telling* you, it was awful."

For those patients who were near the end, the nurses would not let them die alone. "Hearing is the last of the senses to go, so even if they were unconscious, we always talked to our GI patients and held their hand or touched them in some way," Lynn said.

The nurses worked twelve-hour shifts six days a week, had a day off, and then switched shifts. If fighting tapered off and there weren't many casualties, work slowed down. Hospital staff could take breaks, an hour or two to relax. But mostly they were busy, dealing with at least a dozen new patients every day. They had to stabilize the wounded before there was any hope of getting them out of the country, and that was time-consuming. As soon as someone was stable, he was moved to another ward in the hospital for two or three days, and then evacuated by C-130 to Da Nang.

Lynn had been working there four weeks when, late on the morning of July 10, the hospital got a phone call alerting its staff that casualties from an explosion at nearby LZ Bayonet were on the way. Fifteen minutes later, the wounded arrived.

At first Nicky went elsewhere. A Huey dust-off took him to Chu Lai's other hospital, three miles away. A dozen long white Quonset huts in neat rows set along Highway 1, in the upper end of the base just north of the main gate, made up the 27th Surgical Hospital. It had thirty doctors with the latest equipment to handle severe trauma cases. From its helipad, Nicky was carried on a stretcher into the emergency room.

According to clinical records I obtained from the National Personnel Records Center, a nurse scribbled on her pad:

10 July 69: 1050—Received in ER via dustoff...

1100—X-rays done.

1115—Received in pre-op. Wounds prepped for surgery.

Major Alton Gross, an orthopedic surgeon, saw that Nicky had multiple fragment wounds to his legs, thighs, forearms and right hand, the femur in his left leg was shattered, and the lower part of the leg was in shreds. Nicky had to have surgery immediately. The operation began at noon and lasted two-and-a-half hours. Gross, assisted by Dr. Bradley Billington, cut away the dead tissue from Nicky's fragment wounds, aligned the fracture of the femur, pinned it and applied a long cast, and amputated the tattered left leg below the knee. Afterward, Nicky was listed as stable.

The next day, Gross filled out paperwork to have Nicky flown out of the country within three days and transferred to another hospital for "FX [fracture] healing and prosthetic fitting." He would have to be hospitalized for at least six months, maybe as long as a year.

Nicky seemed to be stabilizing; he was conscious and aware of his surroundings. But that didn't last. At noon, a nurse noted that his condition was deteriorating. He was "very disoriented, confused as to time and place," she wrote.

Gross went to see why his patient in Bed 4 was having trouble. He found that Nicky was losing consciousness, his respiratory rate had shot upward and his heart was straining. He was feverish, with a temperature of 102. Tests showed the oxygen level in his blood was alarmingly low. Gross suspected pulmonary emboli, which can occur after injury to a long bone. Microscopic droplets of fat probably had spilled from the bone marrow of Nicky's shattered femur—the longest bone in the human body—and clogged blood vessels in his chest. As a result, his lungs were struggling to aerate his blood. A urinalysis and spinal tap were done in a search for telltale fat droplets. The tests on the urine and cerebrospinal fluid came up negative, but that didn't necessarily rule out the diagnosis.

Nicky was in no condition to be flown out of the country, so Gross dropped the idea. With a phone call, he arranged for his patient's transfer to the evacuation hospital. At 2:30 that afternoon, as Nicky was being readied for the move, he was given an intravenous push of the anticoagulant heparin. He was listed as very seriously ill at five that afternoon when admitted to the 312th Evac.

Tony Viall, Tom Sled and Billy Vachon had been transported there the day before. Billy was rushed into surgery. Tony and Tom were placed on stretchers and brought inside, where the stretchers were laid on sawhorse-type frames. Lying side by side, they made nervous jokes and felt pain as corpsmen cut off their clothing and shaved and scrubbed them, touching and rubbing against their wounds. An older nurse came by and asked if she could do anything for them. She gave them cigarettes and lit them.

Tom yelped when a medic with forceps yanked a chunk of the grenade's iron casing out of his elbow, not realizing how deeply it was embedded in muscle and bone. Aware that his thigh was bleeding, Tom asked about his "jewels" and a nurse said, "They look good to me, sir." Then he passed out.

Tony had surgery to remove metal fragments from below the knee in his left leg, which appeared so bloodied in the moments after the blast that he was afraid he'd lose it. Another fragment had pierced his right foot. He figured the piece had ricocheted off his steel helmet, which he'd placed on the floor after sitting on the bench.

In recovery, Tony asked about Nicky. "He lost a leg," said a helicopter crew chief who was visiting someone in the hospital. Tony was stunned. It was a terrible thing to lose a leg. At least his friend was alive and would be going home. As for Billy, who had been seated beside Nicky when the grenade went off but hadn't bled profusely, Tony assumed he would be all right and didn't ask about him.

⌒

The day after his friends were flown to the 312th Evac, Nicky was in the hospital's intensive care unit with a killer snaking through his bloodstream—the pulmonary emboli. There, Nicky would cross paths with Second Lieutenant Lynn O'Malley of the Army Nurse Corps. Billy Vachon occupied one of the other beds in the room.

The records of Nicky's care are incomplete, but they do show that on July 12 he was in guarded condition. That day, his parents got a telegram from the Army.

It was a Saturday, late in the afternoon. Uncle Louie and Aunt Bert were watching TV in their living room when Louie saw a deliveryman walking across the front yard and went out to meet him. They talked, and Louie signed for a Western Union telegram. He tore open the flap on the pale yellow envelope. Inside were two half-pages of teletype text. His eyes raced over the words.

> The secretary of the Army has asked me to express his deep regret that your son ... was injured in Vietnam on 10 July 1969 by fragments while in a training session when a grenade accidentally detonated. He received wounds to both of his legs with resultant surgical amputation of his left leg below the knee and a fracture of the femur of his left leg and wounds to his right hand. He has possible fat microemboli to the lungs and brain and possible pulmonary embolus. On 11 July 1969 he was placed on the very seriously ill list and in the judgment of the attending physician, his condition is of such severity that there is cause for concern.

The Western Union telegram to Louie Venditti informing him on July 12, 1969, that his son Nicky had been seriously wounded two days earlier in Vietnam "while in a training session when a grenade accidentally detonated." No more word came from the Army until after Nicky's death.

Thunderstruck, Louie could hardly believe it. He had just seen Nicky here at home. His son had just gotten to Vietnam. Now he's lost a leg and might die!

There was no way to know how Nicky was doing at that moment, in a military hospital 8,500 miles away, unreachable. Louie and the rest of the family and their friends remained at the mercy of the Army, and no further word came. The hardest thing was the helplessness of not knowing. After a few days, Louie thought: Why not call the president? Surely, Nixon could get someone to pass information about Nicky's condition to the family. Louie managed to reach someone at the White House, but nothing came of it. There was nothing to do but wait.

On the evening of the Saturday that the telegram came, Nicky's fiancée, Terri Pezick, had a baby-sitting job. It was still early when the couple who hired her returned. As they were driving her home on King Road, they passed Louie and Bert's house. Terri noticed more than the usual number of cars parked there.

Oh, looks like they have company, she thought.

When she got home, her stepfather broke the news that Nicky had been hurt. He tried to soften the blow by saying he didn't believe he was seriously wounded, but that didn't help. Her mind went blank. The parked cars she had seen at Louie and Bert's belonged to friends and relatives who had come to support the couple as they tried to adjust to this grim reality and as they waited for some snippet of new information about Nicky that would ease their minds.

Word of the telegram zipped around little Malvern, whose residents were shaken by the misfortune that had befallen a hometown boy. The extent of Nicky's injuries wasn't clear to many. Louie might have been telling people his son wasn't in great danger, reflecting his own need to deny the reality. Others might have deliberately understated the case to shield people who were close to Nicky.

Mary Anne Wallace was at her accounting job when her dad called to say Nicky had been wounded.

"How bad?" she asked.

"Well, I don't think too bad," her dad said.

Even if Nicky were seriously hurt, Mary Anne thought he would pull through. He had always been tough and a fighter. She prayed for him, held on to her faith in his recovery, and got up the courage that day to call her adversary, Terri, to pitch a truce. But Terri wasn't home.

A few days later at work, Mary Anne dialed Terri's number again.

This time Terri answered. She was overwrought with worry, and for the first time, Mary Anne understood how strongly Terri felt about Nicky.

"I know you probably don't want to talk to me," Mary Anne began.

Terri didn't stop her. Mary Anne said it was wrong for two people who cared about Nicky to be at odds with each other, and that when Nicky came home, he would need all the support he could get. "It would make him happy to know we are together," she said.

At the time, Mary Anne was busy getting ready for her wedding, which would be held in September. Terri agreed that for Nicky's sake, they should be friends.

Suspense tugged at everyone. Louie and Bert got no updates from the Army on Nicky's condition after the initial telegram, and they hoped that meant Nicky was recovering. But each hour that he lay in the evacuation hospital at Chu Lai, Nicky was slipping closer to death. From the first examination of his patient, thoracic surgeon Samir Marrash knew that time was running out. Nicky couldn't breathe. His lungs were oxygenating poorly and filling with fluid. They also might have been damaged by the grenade's concussion wave in the closed classroom. Nicky had a fever for several days, indicating an infection. When he got blood transfusions, the blood wasn't clotting properly to control hemorrhaging—a process necessary for him to survive.

Marrash was a thirty-six-year-old Army major born to Syrian parents in Sudan and a graduate of the medical school at American University of Beirut. His diagnosis was the same as Gross's: pulmonary emboli.

Marrash did not remember Nicky three decades after serving in Vietnam, but he believed Nicky was probably on a ventilator. The machine helped him breathe but couldn't stop the fatty emboli from clogging his bloodstream. Records show X-rays were taken and Nicky's blood was tested regularly. They also show that on July 14, at four o'clock in the morning, Lynn O'Malley gave Nicky 500 milliliters of whole blood in a process that lasted two hours. At 6:15 a.m., she administered another 500 milliliters.

Lynn, like the doctor, does not remember Nicky. Too many young men, horribly wounded, came through the doors of the ICU and the adjacent recovery room, blurring the memories of the medical personnel who tried to save them.

At 4:15 p.m. on Tuesday, July 15, the day President Nixon sent a letter to North Vietnamese leader Ho Chi Minh that helped start secret peace talks in Paris, Nicky died. His body was sent north to the Army Mortuary at Da Nang.

Uncle Louie was at work at Foote Mineral on Wednesday, July 16, and Aunt Bert was having coffee in her kitchen with a friend, police Chief Cockerham's wife, when two men in Army uniforms came to the door. Bert immediately knew why they had come, and slammed the door in their faces before anyone said a word. She rushed to the kitchen closet, squeezed herself in and pulled the door shut. She wouldn't listen to anyone who tried to talk her out. Eventually, someone half-dragged her.

The young soldiers brought a Western Union telegram from Washington: "The Secretary of the Army has asked me to express his deep regret that your son ... died in Vietnam on 15 July 1969."

I was home alone when the phone rang. It was Uncle Louie. His voice was flat, and he was crying. He wanted to talk with his younger brother, but Dad was at his bookkeeping job.

"Tell your father," Louie said, then he paused and his voice quavered, "that Nicky is dead."

"I will." That was all I could say.

Someone called my mom, who was at her job at an electronics company near Malvern, and said she should get right over to Louie and Bert's. When she arrived, she found Louie wandering around the house in a fog. Bert had retreated into their bedroom and, as she had done in the kitchen, closed the door. Mom found things to do—washing dishes, running the sweeper, tidying up. Occasionally she went to check the bedroom door but didn't knock because she didn't want to intrude. She heard nothing from inside, and Bert didn't come out.

A few days later in Portland, Maine, Billy Vachon's mother looked out her kitchen window and saw a brown Army car coming down the street toward the house. Panic seized her. She rushed across the room to a picture window that looked out on Casco Bay just beyond her backyard, and wanted more than anything to run along the water's edge, never stopping and never turning back.

Unlike mine, Billy's family had received multiple telegrams from the Army with updates on his deteriorating condition. The twenty-one-year-old pilot had been admitted to the 312th Evac at 10:40 a.m., less than half an hour after the explosion. Clinical records show that metal fragments had hit him in the left front of his skull, fracturing it and reaching his brain. Fragments had also penetrated his groin, buttocks, both arms and both legs.

Army doctor Harry C. Sherman performed a craniectomy, excising the damaged part of Billy's skull, and cut away the dead tissue from Billy's brain and his other wounds to prevent infection.

At 9:40 the night of July 10, Billy was classified as seriously ill but expected to recover. In the ensuing hours, however, his body couldn't cope with the damage. Three days later, Lieutenant Colonel Sherman listed Billy as not likely to recover. Then on July 16, Sherman amputated both of Billy's legs below the knee. In Maine, Billy's father, a World War II combat veteran who retired from the Army in 1965 as a sergeant first class, prepared to board a plane to Vietnam to see his son. U.S. Senator Margaret Chase Smith, a Maine Republican, helped set up the trip

Billy Vachon, twenty-one, was mortally wounded July 10, 1969, in the explosion that rocked the orientation building at LZ Bayonet. He died seven days later in the intensive care unit at the 312th Evacuation Hospital, Chu Lai.

after learning what happened to Billy from someone outside the family. But at 3:25 on the morning of the seventeenth, Billy died.

Wilbur Joseph Vachon III, a husband and father of a two-year-old girl, had outlived Nicky by two days. A certificate listed the cause of death as "multiple fragment wounds to head, extremities and pelvis."

Nicky and Billy were dead, and Tony Viall and Tom Sled were badly hurt. Tom was patched up and sent to the convalescent center at Cam Ranh Bay for two weeks, then to Japan for surgery aimed at reconnecting the shredded peroneal nerve at his knee. He stayed there for about six weeks, watching Japanese-dubbed re-runs of *Zorro* and *Bonanza* on television. At Travis Air Force Base, he had two more surgeries, followed by months of post-operative care and therapy. But, he said, "the nerve never took and they gave up on it." He was rated fifty percent disabled and took a medical retirement.

Tony also was sent to the convalescent center at Cam Ranh Bay, where he wrote to his older brother that Nicky had lost a leg and was probably on his way home. But his brother, an Air Force veteran in

Florida, already knew what had happened. He had heard about it from
his mother, who had heard it from Skip Smith's mother, also in Rossville. Elsie Smith and Bert had become jovial, fast friends at Fort Rucker
when their sons got their wings and had kept up a Georgia-Pennsylvania
correspondence. Tony's brother wrote back with the news that Nicky
was dead.

Tony couldn't believe it. Yes, his buddy had gone into shock and
lost a leg, but people lose legs and survive. All along Tony had been
thinking: That lucky devil, he's in a hospital in Japan, waiting to be air-
evac'd to a hospital in the States.

Later, Tony learned that Billy too had died. It was another gut-
wrenching blow.

When Tony was released from the hospital at Cam Ranh Bay, he
returned to Chu Lai. He tried to find out why the grenade had gone off
during the orientation class. Was it an accident or intentional? The
replacement company commander told him only that the instructor
had survived the blast and suffered a breakdown.

At American Division headquarters, Tony was assigned to Company A of the 123rd Aviation Battalion and flew with that unit. But his
fragment-pierced right foot gave him increasingly more pain, and it
affected his walking, so the flight surgeon took him back to the 312th
Evac. His foot had become infected and he would need a skin graft. As
a result, Tony's tour in Vietnam was over. He was flown to Japan for
the operation, then to Hawaii, then California, and on September 28,
he arrived home in Rossville. Elsie Smith wrote to Bert two days later:
"He seems to be in real good condition, except for a slight limp. He
will be home until Oct. 17th, then he reports to Ft. Rucker for 16 weeks.
After that he don't know what his orders will be. I'm very glad he don't
have to go to a hospital. He came to see me yesterday and talked about
what happened. He is very angry and bitter about it, and so are we."

His anger stayed with him.

"Something that tragic takes a while to actually sink in," he told me.
"It left me with an empty feeling. I wanted to get away from all of it."

At Fort Rucker, he served in the Aviation Armament Division. He
took advantage of an "early out" offer in 1971 and worked for a chemical
company in Chattanooga. But he returned to the Army in 1976 to fly
helicopters at Fort Campbell, Kentucky, with the 101st Airborne Division (Air Assault). In less than two years, he left the service for good.
He had two sons, then grandchildren, then great-grandchildren.

～

Some 560 miles from Rossville and two months before Tony's September 1969 homecoming, an Army specialist fifth class named Timothy T. Williams was laid to rest in an Ohio cemetery. The twenty-three-year-old equipment repairman from the Toledo area died instantly when the grenade went off in the LZ Bayonet classroom, sending a fragment into his chest, mutilating two fingers and slashing his face, left arm and both legs. Assigned as a replacement to the headquarters company of the 26th Engineer Battalion, he was just a few days into a one-year tour.

Tim was not in the Army by choice. At Rossford High School he bowled, played drums in the school band, escorted the homecoming queen and won honorable mention in the National Merit Scholar exams. After graduating in 1963, he wasn't interested in college; instead he worked as a welder at the local Chrysler plant. Drafted in 1967, he first served a year in a quartermaster battalion in Germany.

Timothy T. Williams of Toledo, Ohio, was drafted into the Army in 1967 and served in Germany before going to Vietnam. A specialist fifth class with the 26th Engineer Battalion, he was killed instantly in the explosion in the orientation building at LZ Bayonet. He was twenty-three.

His older brother Gary, an Air Force veteran, told me that Tim was persuaded to re-enlist on a promise that he would continue to be stationed in Germany, but a week later he got orders to Vietnam. Just before he left for the war, he asked his girlfriend to marry him. She said the timing wasn't right.

On July 10, 1969, Tim's body was taken from LZ Bayonet to the 27th Surgical Hospital and examined by Dr. Billington, the surgeon who helped amputate Nicky's leg that day. Tim was sent on to Da Nang and flown to Travis Air Force Base. From there, an Army escort accompanied his body to Rossford, Ohio. Tim's funeral in nearby Toledo Memorial Park took place on July 21, the day after the Apollo 11 moon landing. "How odd that seemed," his sister, Jill, wrote to me, "that our government could be advancing so quickly in conquering worlds beyond our own, while it couldn't find a way to bring its own sons home from a third-world country of horror."

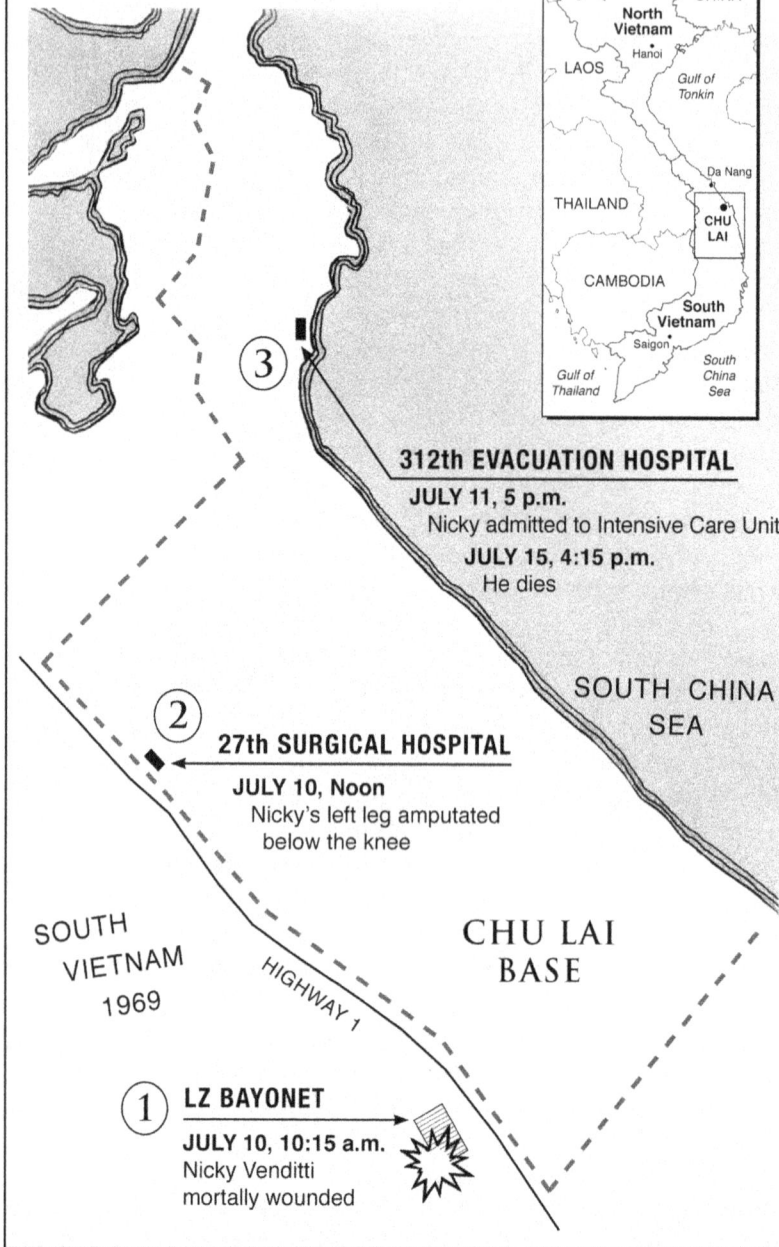

NICKY'S DEADLY PATH

312th EVACUATION HOSPITAL

JULY 11, 5 p.m.
Nicky admitted to Intensive Care Unit

JULY 15, 4:15 p.m.
He dies

SOUTH CHINA SEA

27th SURGICAL HOSPITAL

JULY 10, Noon
Nicky's left leg amputated
below the knee

SOUTH VIETNAM 1969

HIGHWAY 1

CHU LAI BASE

LZ BAYONET

JULY 10, 10:15 a.m.
Nicky Venditti
mortally wounded

CHINA

North Vietnam

Hanoi

LAOS

Gulf of Tonkin

THAILAND

Da Nang

CHU LAI

CAMBODIA

South Vietnam

Saigon

Gulf of Thailand

South China Sea

Warrant Officer Nicholas L. Venditti arrived in Vietnam on the Fourth of July, 1969. He was undergoing American Division orientation six days later when an explosion cut him down. Map by Kenneth F. Raniere.

The grenade had claimed three lives. Tim was the only one killed outright. Nicky died five days after the explosion. Billy died seven days after it. Both pilots had been under intensive care by Army doctors and nurses who, in their day-to-day jobs, saw a ceaseless parade of suffering and untimely death.

For those like Lynn O'Malley, as with the men who fought in the war, the trauma of Vietnam followed them home. In 1970, Lynn returned to eastern Pennsylvania, and while stationed at Valley Forge General Hospital, she spent quiet night shifts in the amputee ward looking through patients' old medical records in hopes of finding someone she had cared for, so she could know that she had made a difference. She found five such patients and told them she had been one of the nurses who tended to them in Vietnam.

Lynn remained in government service, ultimately becoming the nurse manager at the Allentown Veterans Affairs Outpatient Clinic. I unexpectedly found her and connected her to Nicky through a project I headed at *The Morning Call*. The newspaper was doing a special section for Veterans Day 1998, and I asked the reporter to include a woman among the veterans he interviewed. He came back saying he'd found "a great nurse," Lynn Bedics, and with each answer he gave to my questions about her, I grew more excited. Where was she? Chu Lai, Vietnam. When? 1969. Which hospital? The evacuation hospital. What kind of duty? Intensive care.

The reporter had gotten Lynn's maiden name, which matched the "L. O'Malley" on Nicky's clinical records. It was incredible. A nurse who had been at Nicky's side in his final hours worked practically in my back yard. Retired now, she lives a half-dozen blocks from my house.

The day I met with Lynn in her office at the VA clinic, I showed her the paperwork I'd gotten from the National Personnel Records Center in St. Louis—Nicky's blood transfusion reports. She froze. Her mouth hung open.

"That *is* my signature," she said softly, not looking up.

But she did not remember Nicky's name or recognize his face when I showed her some photos of him. She explained: "Most of my patients could not communicate because of sedation, injuries or being on a ventilator. With all those tubes around your face, it would be tough to recognize your best friend. The volume of patients I saw was overwhelming. If I worked 290 days and saw ten new faces a day, that would be 2,900 patients I saw, and I'd bet I saw a lot more than that."

Still, she had been with Nicky. She had stood by his bed, touched him, tried to lessen his agony, might even have spoken words of comfort. When she raised sorrowful eyes to mine, I imagined she might have looked at Nicky the same way over several days in July 1969, knowing there was nothing she or anyone else could do to save him, that it was only a matter of hours or a few days before he succumbed to his injuries.

The following year, and many hundreds of patients later, Lynn came home to Walnutport in Pennsylvania's Lehigh Valley. Ahead of her was her ongoing work at Valley Forge with the wounded of the war. But then, right after her return from Vietnam, all she wanted to do was forget. Her mother had saved all the letters Lynn wrote while she was in Vietnam and gave them to her. Lynn took the shoebox full of letters into the back yard and burned them.

Chapter 10

Casting a Wide Net:
1994–2015

In the fall of 1994, I knew nothing of any of this—no names, no details, nothing that would allow me to envision what had happened to my cousin. All I had at the start was the rudimentary truth of Nicky's fate. But I assumed it was just a matter of months before I'd know the whole story of what had happened that fiercely hot morning of July 10, 1969, when Nicky and his friends hopped off trucks at LZ Bayonet, sat up front in the orientation building and watched as an instructor let loose a grenade that rolled to a stop under their table.

Now I knew that Nicky didn't die on his eleventh day in Vietnam as a result of an enemy rocket attack while waiting with other guys for a transport, the story I had heard as a boy and long held onto as an understandable end befitting a proud soldier. No, the disturbing reality was that he died at the hands of another American, a soldier holding a class on grenades in the supposed safety of a landing zone that was home to an entire U.S. infantry brigade.

Nicky, Billy Vachon and Tim Williams all dead from the explosion, Tony Viall and Tom Sled badly hurt and many others injured among the several dozen in the classroom. I was sure that such an extraordinary occurrence—horrifying, outrageous, an appalling waste of life— would have generated an Army record stating exactly what happened, how it happened, the total number of casualties and who was held responsible.

I had to know because, from that moment in 1994, Nicky had become more than the son of my dad's brother, one of many cousins. He had become a part of me in a melding richer than blood. And though his life had ended violently before he ever had a chance to share it with me, I was now reaching across a chasm of time and space to find him.

As I had seen in my 1995 visit with Uncle Louie and Aunt Bert, they still ached over their loss. So, too, I would find in the months ahead, did the families of Billy Vachon and Tim Williams. All had been torn apart. I thought that with the quest I had undertaken, maybe I could help them and myself. The Army had treated all three families callously in 1969, giving only this curt explanation: Nicky, Billy and Tim were in a classroom when an instructor accidentally detonated a grenade. The families deserved to know the complete story, and I had the tools and determination to accomplish that.

All I had to do, I thought, was get the documents. The Army surely would have a scrupulous account of what happened in that classroom at LZ Bayonet, so highly visible because it lay just off a major U.S. base, the headquarters of the largest Army division in Vietnam. The answers would be there, the mystery solved. Perhaps in mere months, I would unearth this narrative from an obscure file cabinet in a government repository. I would find a paper trail of revelation. The story would have a beginning, a middle and an end.

I should have known never to assume.

I started with a document I'd gotten in February 1995, only a few months after speaking with Doug Howard, the Army mortuary program specialist who had told me what really happened to Nicky and set me off on this venture. The document was Nicky's Individual Deceased Personnel File from the Army's Mortuary Affairs and Casualty Support Division in Alexandria, Virginia. Among pages and pages on the disposition of his remains and property, it says only that he died after he was "injured while in classroom at base camp when the instructor of class accidentally detonated live grenade," the jolt that Howard had delivered to me.

A document perhaps more likely to paint a bigger picture would be Nicky's Official Military Personnel File, so I requested a copy from the National Personnel Records Center in St. Louis. But that, too, came up frustratingly short. In its twenty-one pages, there was just a one-page Statement of Medical Examination and Duty Status that had any detail—and it is sparse: "During a period of classroom instruction, instructor unknowingly discharged a live grenade at approximately 1015 hours 10 July 1969, LZ Bayonet, RVN [Republic of Vietnam]."

That's it. No mention of anyone else, whether anyone besides Nicky was killed or seriously injured and who they were, who the instructor was and what had happened to him, whether there had been an investigation and, if so, what were the findings. The two files on Nicky just focused on him, and that was all I had by the time I sat down with Uncle Louie and Aunt Bert in early December 1995. Unsure of what they might have preserved, and eager to know, I was excited to find that they had saved everything they'd gotten from the Army— telegrams and letters, all still in their envelopes, including form-letter condolences from the Americal Division commander, the division chaplain and the secretary of the Army, even a note from a Chu Lai hospital chaplain who said he'd given Nicky last rites and that a Mass had been said for him and his family. Louie and Bert turned all of these things over to me.

One of the condolence letters had an explanation. The letter to Uncle Louie was dated August 14, 1969, a month after the incident, and came from the headquarters of the 23rd Adjutant General Replacement Company, the unit at Chu Lai through which newly arrived troops passed on the way to their assignments.

> On the morning of July 10, 1969, Nicholas was attending a class on the use of grenades at the Americal Division Combat Center.... At 10:15 a.m., the class instructor removed the safety pin from a hand grenade that was thought to have been disarmed for instructional purposes. However, the grenade detonated when he threw it to the floor of the classroom.

What hit me right away was that the instructor hadn't set off the grenade while doing something casual, like picking it up or setting it down, or demonstrating how it could be disarmed. One might think that was the case from the wording in the telegrams Uncle Louie received earlier—messages that simply said the instructor had accidentally detonated a live grenade.

No, this letter painted a different, more disturbing picture: He hadn't merely fumbled the grenade or mishandled it. He had pulled the pin and thrown it. That sounded reckless to me.

Presumably, he had given his lecture before without incident. Unless he had suddenly come unhinged and wanted to kill himself and take some fellow soldiers with him, he didn't know the grenade in his hand would explode. Apparently, he had a real grenade, one that would go off if it were filled with explosive and its blasting mechanism was intact.

Why tempt fate in a demonstration? Even if he had rendered it inert, why use a real grenade rather than a dummy clearly marked for practice? If the idea were to make his presentation realistic, to ensure he had the rapt attention of the men in front of him, it sounded to me like a foolish, deadly gambit. On any given day, he might unwittingly pick up the wrong prop, a live fragmentation grenade, and not know until it was too late. Or someone, in a calculation meant to maim and kill, might switch his grenades.

There was an unnerving irony to it. Nicky died as a result of a safety class. And though the blood was on this sergeant's hands, it was possible he wasn't the only one to blame for what I certainly thought was a shocking lack of common sense. It could be that someone above him in the chain of command allowed, encouraged or even ordered him to roll a grenade in the class, making an accident possible.

And that's another thing I noticed about the letter from the replacement company: Even though it says the grenade "was thought to have been disarmed," it doesn't contain the words accident or accidentally. Nor does it mention an investigation.

A few weeks before my meeting with Uncle Louie and Aunt Bert, I had written to the National Personnel Records Center in search of an incident report. The response pointed out that reports on accidents are destroyed after six years. That really set me back. It was such a disappointment. Later, a veteran who had flown helicopters in Vietnam suggested I try the place that holds reports on Army accidents, the U.S. Army Combat Readiness/Safety Center at Fort Rucker, Alabama. Again I got nowhere. Its ground accident reports, a lieutenant colonel there informed me, only date to September 1973. Before that, he wrote, accident reports were kept on file at local safety offices "for a period of time" and then destroyed.

This was rotten luck for me, but I wouldn't allow myself to give up. I wondered if what happened might have been written up and filed as something other than an accident report. Or maybe the information in such a report had been copied to another file that was stored somewhere. When I sought World War II records of my Uncle Sam, his Army records didn't exist—apparently they burned in the 1973 fire at the National Personnel Records Center that destroyed millions of military personnel files. But information that would have been in Sam's records in St. Louis was also in his file at the Department of Veterans Affairs because of the medical care he'd been given, and I was able to get that file from the VA's regional office in Philadelphia.

There was something else, too. It just seemed to me that the deaths of several soldiers from an Army safety class on grenades would have been treated as more than an accident. It seemed so unusual, so consequential, so much more deeply serious than a jeep hitting a tree or a helicopter crashing from mechanical failure—the kind of occurrences that along with disease, drug abuse, homicide, suicide and friendly fire accounted for almost a fifth of the 58,220 American deaths in Vietnam. And sure enough as I pressed ahead, I found support for this hunch when some experts suggested there would have been a more extensive paper trail.

This drew me inexorably into an Alice-in-Wonderland vortex of military bureaucracy as one agency after another pointed elsewhere for answers. The National Personnel Records Center steered me to the National Archives and Records Administration. But in February 1996, the archives responded that it was "unable to locate records of this incident." Further on, my requests for Army records under the Freedom of Information Act turned up nothing, not even an Americal Division report under the Army regulation for investigating "serious incidents," called an AR 15-6 investigation. In this scenario, a commander assigns an officer to conduct an investigation and write a report, which can be used to formally charge someone deemed responsible or to drop any further action.

I'd also written to the Army Center of Military History in Washington and gotten a response in November 1995 recommending that I seek the Operational Reports/Lessons Learned for the Americal Division and the 16th Combat Aviation Group, which was attached to the division. Right away, I ordered those documents from the Army Military History Institute at Carlisle, Pennsylvania. The Americal Division's report for the quarter ending July 31, 1969, is a once-classified, seventy-four page document that lists changes in command, combat operations, death tolls on both sides and other matters such as those dealing with the troops' morale—attendance at religious services and complaints of tardy mail delivery. I thought I'd hit pay dirt, just from the title. If the explosion at LZ Bayonet wasn't a lesson learned, what was it?

Yet there is no mention of it. One section simply states the Americal Division "continued the mission of conducting in-country orientation and replacement training." But another part does offer this key point: "During the reporting period, there were no reportable incidents of known sabotage, subversion or espionage."

The corresponding report for the 16th Combat Aviation Group, to which Nicky was assigned as a helicopter pilot, reveals nothing about his fate or any of the others in the training room that day.

This was getting harder. It grated on me that a detailed picture of what happened remained elusive no matter where I turned. How bad was it? How many casualties? I knew from Aunt Bert that Billy Vachon had died from the blast, and I had confirmed that through online sources and contact with Billy's family in Maine. But had anyone else been killed?

I discovered there was a painstaking way to find out about casualties. The National Archives' Center for Electronic Records keeps casualty records compiled by the Defense Department and the Army adjutant general's office. They explicitly state how a person died and where. But the documents are coded and hard for a layman to read without help. So through an exchange of emails, archives specialist Ted Hull patiently answered my questions and explained how to read the codes. He said that with what I knew about the incident, I might be able to "piece together other names." After this coaching, I requested the records of Americal Division deaths on July 10, 1969, the date of the explosion, and for the next ten days. A packet of records printed from a computer database came in the mail, pages of names listed in rows and followed by a jumble of letters and numbers. I pored over the pages, matching the codes to a key. It was tedious but eventually I got my answer.

Nicky and Billy Vachon were both listed as "non-battle dead," with "grenades" under the heading Complementary Cause. One other man fit the same descriptions exactly, Specialist 5 Timothy T. Williams, listed as having died on the day of the explosion. I called Doug Howard, the mortuary program specialist who'd been so helpful. He confirmed that Williams died in the classroom and that there were no other deaths from the explosion.

Three dead—it was satisfying to at least nail that down. It was a sign that if I couldn't get one document that had everything I wanted to know, maybe a piecemeal approach would work. Maybe I could keep pulling information from different sources. The next step was getting Billy's and Tim's service and casualty files, the same records I had for Nicky. That was easy because, as I had found in seeking Nicky's paperwork, a person's military service is a matter of public record, though personal information is not.

A handwritten entry in Tim's casualty file got my attention because it indicates there was an investigation: "In classroom at LZ when instructor of class accidently detonated a live grenade. Investigation is in progress." The file lists the funeral home in Rossford, Ohio, that handled the arrangements for his burial. I called and got the names of family members, found them and contacted Tim's stepmother and his sister and two brothers. They had never gotten a complete accounting from the Army of what happened to Tim.

I kept chipping away, sometimes to no avail.

A few scattered lines did turn up here and there, thanks to the National Archives. One of these documents was the daily staff journal of the Americal Division's 198th Light Infantry Brigade, based at LZ Bayonet. An entry made at 11 a.m. July 10 at the brigade's Tactical Operations Center says an M26 fragmentation grenade exploded during a Combat Center class on the LZ, injuring twenty-five Americans and two South Vietnamese of the local Popular Forces. That's all there is to it. But the entry was made just forty-five minutes after the grenade went off and was therefore only preliminary. It doesn't mention, for example, that one man—Tim Williams—was already dead.

Another daily staff journal, the one kept at division's Tactical Operations Center at Chu Lai, has an entry that was made five minutes later. It states the instructor at LZ Bayonet was giving a class on claymore mines, commonly used anti-personnel mines that blast fragments in a fan shape several feet above the ground. He was telling the replacements not to rig a grenade to a claymore. But in trying to demonstrate why it's risky to booby-trap claymores, he somehow used a live M26 fragmentation grenade instead of an inert one. The grenade detonated when the claymore was lifted off it. This journal entry also put the number of wounded at twenty-five, noting the total was verified in a head count by a lieutenant from the Combat Center. The wounded, it said, were all evacuated by 10:35 a.m.

Tony Viall and Tom Sled had not remembered a demonstration involving a claymore mine, only that the instructor lobbed his grenade.

Continuing to reach for anything, I tried to get a list of casualties treated at Chu Lai's two hospitals—the 27th Surgical and the 312th Evac—but both the personnel records center and the National Archives said they didn't have them, with the archives adding that they probably didn't exist.

I did connect on one inquiry, when I asked the National Personnel Records Center to send me copies of the paperwork on Nicky's care at the 27th Surgical and 312th Evac. I didn't even know if the government kept such records. But to my surprise in March 1998, a packet came in the mail with Nicky's clinical records from both hospitals—nearly fifty pages of nurses' notes, doctors' orders, radiographic reports, electrocardiographic records, operation reports, progress reports, lab reports, blood transfusion forms. It led me to Lynn O'Malley Bedics, the Army nurse who tended to Nicky as he lay dying at the 312th Evac. Flush with that find, I got the clinical records for Billy Vachon as well.

The paper search had me mining the Internet and in email contact with Vietnam veterans and others who offered to help. I was learning the names of various reports that existed in the military's labyrinthine bureaucracy. In the 1960s, for example, the Army had Blue Bell Reports on criminal activities and Blue Bonnet Reports on serious accidents—incongruous names for the subjects they dealt with. There were also Serious Incident Reports, which came into play after "a significant incident, crime, accident, wrongdoing or mismanagement" involving Army personnel, property or equipment. Military police, as the Army's law enforcers, conduct investigations, and so do agents of the Army's Criminal Investigation Division. Both the MPs and the CID are overseen by the provost marshal general. A CID Report of Investigation includes a case summary, witness statements and a statement by the investigator.

One of my contacts in the spring of 1997 was a thirty-year-old peacetime Army veteran and Drew University graduate history student who was working on his dissertation on criminal investigation and how military justice was applied during the Vietnam War. George Lepre, a New Jersey resident, would go on to write several military histories, including *Fragging,* a 2011 book that examined why American soldiers attacked their officers in Vietnam. He was just the kind of knowledgeable insider I needed as a guide. He could help me find my way through the maze.

Lepre had this advice: Send a request under the Freedom of Information Act to the Army Crime Records Center at Fort Belvoir, Virginia, the repository for Criminal Investigation Division Reports of Investigation and military police reports. Ask for copies of the Serious Incident Report, the Report of Investigation or the Blue Bell Report. Serious Incident Reports covered significant crimes and accidents, including "murder, voluntary and involuntary manslaughter, assault with intent to commit murder or manslaughter, aggravated assaults, and unlawful or unauthorized discharge of firearms when injury occurs or unfavorable publicity can be expected," which seemed to fit Nicky's case. Such reports were sent to USARV, the Army's top command in South Vietnam.

Fired up about this, and thinking I was really close to linchpin information, I wrote to the center at Fort Belvoir. Its answer was one more disappointment. "A search of the Army criminal file indexes utilizing the information you provided revealed no files responsive to your request." When I appealed, the center came back with the same answer, insisting that "a thorough search" had been made.

Lepre was incredulous. "I can hardly believe that they would not have a report on an incident as grave as this one," he said. "Several helicopter pilots being killed and injured in a grenade explosion is pretty darn serious." He suggested that for the next step, I get in touch with Rich Boylan at the National Archives in College Park, Maryland. Boylan served in Vietnam in 1970, Lepre said, and is "the Vietnam records guru at the National Archives."

I wrote to Boylan, and he got back to me with another dose of dismaying news: There was nothing about what happened to Nicky in American Division Provost Marshal files, which contain the reports of military police investigations. Nothing in the division's Inspector General files, which deal with investigations of misconduct and cases of soldiers killed or injured by friendly fire. And nothing in the Inspector General files for both the Military Assistance Command, Vietnam, which was the command structure for U.S. forces, and USARV. He said the Serious Incident Reports in the National Archives' custody are incomplete and don't cover September 1968 to January 1972, and the Blue Bell Reports are even more incomplete, covering only 1970.

Was it possible that no record existed?

In what might have been a stretch but was worth trying, I even wrote to the Navy. The American Division came under the authority of the 3rd Marine Amphibious Force, the Marine headquarters in Vietnam. It oversaw U.S. military activity in the I Corps tactical zone that included Chu Lai. The Marine Corps is under the Department of the Navy. Maybe the American Division had sent a report to its Marine bosses.

No such luck. In response to my query in 2014, the Navy's Office of the Judge Advocate General at the Washington Navy Yard told me it had nothing on the grenade blast.

Army veterans who had firsthand experience in Vietnam investigations tried to help me. Their emails ricocheted as they discussed the case, and then landed in my "in" box. But they only widened the quagmire.

Al Grande, who became Serious Incident Reports officer in the Office of the Provost Marshal General of the Army in 1972, offered little hope that I'd find what I was looking for. "Since the incident took place in July 1969 and appears to be a fatal training accident, my guess is it met Blue Bonnet reporting criteria," he said, but added, "I seriously doubt the records are still available." Still, he wouldn't say they didn't exist, noting the information might be "sitting in a box at some records holding area somewhere in the universe."

Another distinguished veteran said it was possible that no report of any kind was ever filed, given that LZ Bayonet was in a combat zone—just west of it lay the Rocket Pocket, foothills from which the Viet Cong launched rockets at American Division targets—and the war was raging. "In combat areas, reporting requirements became lax, so there is no absolute assurance that a report was ever sent," said Frank Cohn, who in 1969 was a lieutenant colonel in the USARV Provost Marshal Office at Long Binh. "The problem in this case is that no criminal act appeared to have been verified, so in my mind it is doubtful that a completed Report of Investigation was ever produced."

A key Criminal Investigation Division investigator of the My Lai massacre told me that if the explosion had clearly been an accident, the CID detachment at Chu Lai would not have been involved. Instead, the local commander would have investigated the circumstances and reported to his superiors. If, however, there was reason to believe that the grenade was exchanged for some reason, Andre C.R. Feher said, the local criminal investigators would have stepped in. "I am sure that if it had been decided that the grenades were intentionally switched, that the CID would have opened a case and they would have mentioned something to me while I was working [on the My Lai case] in their Chu Lai office."

Feher checked with the former 8th MP Group CID commander who was in Saigon at the time. "If a CID investigation had been initiated, I am sure he would have remembered. He did not, and if there is no CID report in the Army Crime Records Center, then the CID did not investigate it."

I wondered how a local commander would have been able to determine exactly what happened. Wasn't this deadly event so chilling and out of the ordinary that a CID probe would have been prudent?

Boylan, the Vietnam expert at the National Archives who helped me for more than a decade, shared that opinion as early as the summer of 1998. "The incident should have been investigated to determine whether the deaths and injuries were the consequence of a criminal act or of negligence," he wrote, and the Army Crime Records Center should have the report.

The chief CID investigator of the My Lai killings, which happened in March 1968 near Chu Lai and involved troops from the American

Division, had a similar reaction and was even more pointed. "Even if it was an accident," Bob Zaza thought, "the case would still have been a negligent homicide. There should have been a full CID investigation and a report."

If one existed, though, I couldn't find it anywhere. My early confidence that I could get all of the answers in a matter of months had dissolved into anger, bewilderment and frustration.

It just didn't make sense.

From what I had learned, the explosion sounded serious enough to me that it should have been investigated by an authority other than a local commander. And I agreed with Zaza that a report should have been compiled by the Criminal Investigation Division regardless of whether the explosion was found to be an accident. That report would have clearly laid out what the investigators determined happened, giving a rundown of the instructor's routine, saying how many men were killed or wounded and who they were. It would have named the instructor and given his background and qualifications. It would have presented the results of interviews with him, the other survivors of the blast and the commanders in charge. It would have stated conclusions and given recommendations on discipline, if called for, and how to prevent something like that from happening again. And it would have made its way up the Army's hierarchy, where the brass would review it for possible further action.

One thing that did make sense as I learned the ins and outs of military communication had to do with the practice of notifying next of kin after a death: It was common that the Army never gave the families of Nicky, Billy and Tim a detailed account of what happened. I read about the procedure for informing families of slain soldiers in C.D.B. Bryan's book *Friendly Fire*, which in 1979 became a TV movie of the same name starring Carol Burnett. The true story tells how an Iowa farm couple confronted the Army after their son, a sergeant in the Americal Division, was killed near Chu Lai in February 1970 by American artillery fire. Peg and Gene Mullen demanded to know more about how their son Michael died, but hit a bureaucratic wall that frustrated and enraged them. Bryan quotes from a letter by a Pentagon official who wrote regarding the Mullen case that "under established procedures," a deceased soldier's unit commander writes a letter of sympathy to the next of kin. Other than a death notification, that was all a family would get—a letter that merely touched on the circumstances in a few sentences at most, like the one Uncle Louie got.

Lepre also noted the process in his book *Fragging*. In his research he had obtained many Individual Deceased Personnel Files, which include copies of telegrams and letters sent to relatives. In most if not all Vietnam fatalities, Lepre says, next of kin got an official death notification and a condolence letter from the man's unit commander. As a matter of routine, the Army was not more forthcoming.

But a written record of an investigation would be something else entirely, full of substance, and after years of searching with the help of numerous experts, I reluctantly concluded that no such paperwork existed. It's as if the grenade had never gone off, or that the result was of so little consequence that the families of the dead didn't deserve a full vetting of what had altered their lives forever. I found no evidence of a cover-up. But the paucity of documentation, even though an accident report could have been destroyed years ago, made me wonder whether the Army didn't want to grapple with an event that reflected badly on it and the war it was fighting in Vietnam.

Carl Craig, a Criminal Investigation Division liaison officer stationed in Tokyo from 1966 to 1969 who had some contact with investigators assigned to Vietnam, told me that a local commander might have handled the case. "Looking back at that time, and asking individual CID agents about their tours back then, it was common for many unit commanders to maintain jurisdiction over a multitude of incidents within their units." Given the magnitude of the incident, involving deaths, he, too, thought there should have been a full-blown CID investigation. "If the Crime Records Center doesn't have a record, and they have responded to your request, advising you that no record of the incident exists, along with the things you have done, I'm at a loss to advise you of what can be done forty-four years after the fact," he wrote in 2013. It appears the CID was not involved, he said.

Feher was puzzled by the absence of paperwork but not surprised. "I cannot understand that such an incident was not reported to USARV," he said. "But then again, when I investigated the My Lai massacre, I had quite some trouble with finding documents at the HQ of the Americal Division and in Saigon."

Without a comprehensive report, where could I turn? To those in command, whose names appeared in letters and a few documents I'd managed to gather. Even if the narrative didn't exist on paper, maybe it still lived in the minds of these aging men who had borne the responsibility for running Chu Lai. Enervated and empty-handed, I saw them as my next best hope.

Chapter 11

A Crowd of Commanders, November 20, 1998

The thirtyish manager of Charley's Place was going around to the tables, saying hello to his Friday lunchtime customers and heading our way. Eighty-year-old Lloyd Ramsey and his wife, Glenda, had brought Mary and me to this popular suburban Washington restaurant, upscale like its northern Virginia venue and notable partly because it was frequented by CIA employees from the spy agency's headquarters nearby. Lively and casual, with good food, it was one of the Ramseys' favorite spots.

The couple were trim and smartly dressed on this sunny, sixty-degree day in mid–November 1998—Glenda in a wool shirt-waist plaid dress and tan blazer, her light brown hair cut short, and Lloyd in a plaid shirt, blue blazer and gray slacks. Both octogenarians seemed healthy, though Glenda recently had been hospitalized for a heart condition. Lloyd, balding with wisps of white hair on the sides, looked fit from golfing three days a week. He was talkative, Glenda demure and courteous. About an hour earlier, when Mary and I pulled up to the curb in front of their split-level home in McLean, both had come out onto their wide lawn with big smiles to greet us.

Glenda called her husband "Feller," and as we waited for our meal, they explained why. They'd both grown up in the small Kentucky town of Somerset, where Lloyd's father addressed the youngest of his three sons as "the young feller." His brothers shortened it to Feller, and that's how the folks in town knew him. To Glenda, through fifty-seven years of marriage that had made them parents, grandparents and great-grandparents, he had always been Feller—a playful tag you would not necessarily associate with a retired two-star general.

Minutes after our platters arrived, the manager came to our table but barely had a chance to say hi. Lloyd sprang to his feet, leaned across the table, thrust his right hand out and barked with a grin: "Hi, I'm General Lloyd Ramsey!" Taken aback and amused, the manager seemed unimpressed at meeting someone of very high military rank. He offered only a polite "hello" and a quick grasp of Ramsey's hand. Mary and I caught each other's sideways glances, a touch embarrassed for our gracious host.

Out of the military for twenty-four years but still "the general," this self-assured onetime warrior was the reason we had come to McLean. As Major General Ramsey during the Vietnam War, he had commanded the Americal Division at the time of Nicky's death. He had been the man in charge of all the Army troops at Chu Lai.

After lunch, he and Glenda drove us around McLean in their sedan, showing us the entrance to CIA headquarters and Colin Powell's home. Powell, he pointed out, was one of two soldiers destined for greatness who had served under him at Chu Lai, which made Ramsey tremendously proud. Powell went on to become a four-star general and chairman of the Joint Chiefs of Staff. In a few years, he would become President George W. Bush's secretary of state.

The other soldier was H. Norman Schwarzkopf, who had led coalition forces to victory over Iraq in the 1991 Persian Gulf War. He came to Chu Lai at the end of 1969 to command a battalion at LZ Bayonet.

Ramsey reveled in his link to Schwarzkopf, but "Stormin' Norman" apparently didn't share the sentiment. Without naming Ramsey, he bitterly criticized the general's work in Vietnam, writing later that he was appalled at the lack of discipline among Americal troops and the casual resort-like atmosphere that the top brass promoted. In his autobiography, *It Doesn't Take a Hero*, he describes the officers mess as "almost worthy of a Club Med"—a spacious and very pleasant building on a hilltop with a terrific view of the South China Sea. The officers ate off white tablecloths, using china and wineglasses. Staff officers recited poems they'd written about the day's happenings at headquarters, drawing laughter and applause. After dinner, they watched war movies.

Schwarzkopf complained that while the division had many thousands of men risking their lives out in the jungle, "their senior officers ate off fine china and recited cutesy little poems." He recalled his outrage when, on his first trip to the officers mess, staff colonels invited him back for a tea dance. Ramsey told me that after he read that, he protested to Schwarzkopf it wasn't true that headquarters held tea dances.

My first contact with Ramsey had come about a year earlier, part of my inquiries to Chu Lai's commanders. My theory was that they might bring light where the paper trail had done little to illuminate the details of why Nicky died or the consequences of his sacrifice and that of the two others who had perished with him.

So I wrote Ramsey a letter in November 1997 introducing myself and asking him if he remembered the explosion that killed Nicky. A week later, he called. Yes, he said, he remembered. It was horrible, he said, the worst training accident in his nine months at Chu Lai. So, I asked, what happened as a result? He could not tell me anything of what actually had happened in that classroom or give specifics of what followed. Though disappointed, I stayed in touch with him, exchanging occasional letters and phone calls. He said he enjoyed the contact, signing his letters "Lloyd" and eventually inviting Mary and me to visit.

In August 1998, a large envelope came in the mail. It contained not a word about Nicky, but it did contain an autographed, eight-by-ten glossy photo of Ramsey looking tough as Americal commander in Vietnam and a five-page biography, single-spaced, a history of his service, his education, the leadership courses he had taken and a chronology of his promotions. It listed dozens of citations, decorations, badges and medals he received, including some of the highest awards the Army bestows—the Distinguished Service Medal and the Silver Star. Among them was a Purple Heart with four oak leaf clusters presented during World War II to mark the five times he was wounded. It noted his enthusiasm for sports, especially badminton and golf, "as a morale and health builder."

The black-and-white Army photo shows him at fifty-one posing in front of a tangle of Vietnamese jungle brush. He wears fatigues, a wide black belt with a pouch and a handgun holster, and a visor cap bearing the two stars of a major general. He stands erect, his shoulders pulled back, bare arms relaxed at his sides and his jaw set square over a bull neck. He has graying, short-cropped hair, thin lips forming a half-smile, and hawk-like eyes that squint into the sun. In a corner he had written in black marker, "David, with warm regards and best wishes. Lloyd B. Ramsey, ... Americal Division commanding general June 69–Mar 70."

His memory of the tragedy of Nicky's death was incomplete, yet he had made sure I received his portrait and a meticulous rundown of his accomplishments. Still, I could not give up on Ramsey as a potential source of information. So Mary and I accepted his invitation to visit in the hope that asking questions face to face might jog the general's memory. In the den of his home that fall, as Mary and Glenda sat apart from us and chatted on their own, I asked Ramsey again what he remembered about the explosion.

"It was brought to my attention immediately," and he had been very concerned about it, he said. The Marine Corps had overall control of Chu Lai, so he informed the Marine in charge, Lieutenant General Herman Nickerson, Jr., who told him to let his subordinates handle the investigation. He did and was briefed on the result: It was a training accident. The class was on the use of grenades and the instructor unwittingly had a live one. He pulled the pin and the grenade went off. Somehow his grenades got mixed up, but no one could determine the circumstances.

Did Ramsey remember who did the investigation? No.

Was the Criminal Investigation Division involved? "No," he said, "or I would have known about it."

Did he remember who briefed him? No, but he thought it might have been an officer with the 198th Light Infantry Brigade. That Americal unit, responsible for protecting the Chu Lai base, might have overseen the investigation because the accident happened on its home turf, LZ Bayonet.

Did Ramsey ever see a written report of the findings? No, he said, adding that there should have been one.

That was all I was going to get from him.

Ramsey was one of a dozen officers I sought out over the years who were in the chain of command at Chu Lai when Nicky was mortally wounded in the classroom. By title and responsibility, I reasoned from my research with experts, they should have known what happened that day.

I started with Robert C. Bacon, the retired lieutenant colonel whose name was on the condolence letter to Uncle Louie. As commandant of the 23rd Adjutant General Replacement Company, he led the unit through which Nicky and other new arrivals got their orientation and in-country training. I thought Bacon would be the man in the know. So I tracked down his information, wrote him a note with his condolence letter enclosed, then called him in 1996 at his home in South Carolina.

"I didn't sign it," he said.

Oh no, I thought, what *is* this?

He explained that he hadn't joined the replacement company until July 20, five days after Nicky's death. Someone on the staff dropped the letter on his desk, he said. He refused to sign it because the accident didn't happen "on my watch," he didn't know anything about Nicky—who in the usual form-letter fashion is praised for his "high personal standards, conscientious hard work and warm personality"—and the brief description of a grenade's detonating when the instructor threw it to the floor of the classroom wasn't what Bacon had heard. In the interest of expediency, he said, someone must have signed the letter for him.

At the time, Bacon was a West Point graduate on his second tour of duty in Vietnam. Earlier he had served as an adviser to the South Vietnamese and appeared on the cover of *Life* magazine, leading an ARVN patrol through a field of elephant grass. Soon after coming to Chu Lai in 1969, he left his post at the replacement company to lead the 3rd Battalion of the 21st Infantry, 196th Light Infantry Brigade, earning a Silver Star for gallantry. In August of that year, he was at the center of a controversy about whether troops under his command had mutinied in the Song Chang Valley near Chu Lai, as reported by Horst Faas and Peter Arnett on Page 1 of *The New York Times*—a debate that still simmers online.

The account Bacon had heard about the grenade's going off was the one that appeared in the division's daily staff journal entry the morning of the explosion: The instructor was demonstrating that booby-trapping a grenade with a claymore mine is risky, so don't try it. He had done the same many times before, using inert grenades. Somehow he had a live grenade this time, and the result, Bacon said, was a "horrible, unfortunate accident" that was "highly unusual because so many precautions were taken to avoid accidents." He had heard that five to seven men were killed—an incorrect number—and as many as twenty-five wounded.

If he had been in charge earlier, Bacon said, he would have stopped the grenade-tossing routine as too dangerous.

So, according to the man whose name was on the condolence letter, he hadn't signed the letter and he could tell me nothing of any follow-up or consequences. I moved on.

My next attempt was to identify who had immediately preceded Bacon as the replacement company's leader. The answer was in the Americal Division's Operational Report/Lessons Learned, which lists Lieutenant Colonel George R. Underhill.

Now I felt sure I'd found the right man. But when I reached Underhill at his home in Montana, he said he didn't remember the explosion. The date indicated it happened while he was wrapping up his tour of duty in Vietnam, he said. He promised to check his records, and in the meantime I sent him more information. A month later, he wrote: "I am at a total loss as to what to say to you. For the life of me, I cannot recall anything related to this tragedy.... I cannot believe that I would not have remembered the incident, nor would I have failed to immediately go to the scene of the accident as soon as I would have been made aware of it." He said that when he left the training center, he returned to the States. There was no formal change of command, and he never knew who or if anyone took his place. He had no records or access to any records showing when he came and went, but suspected he had left Chu Lai before July 1969 and was still being listed as the Combat Center commander.

I looked at the names on Nicky's paperwork. The report of his death in his personnel file was signed by Captain Edward R. Canady of the 23rd Adjutant General Replacement Company, responsible for processing the new arrivals. Canady would have been the commander Tony Viall spoke with when he returned to Chu Lai from convalescing at Cam Ranh Bay.

This time I connected with someone who remembered, and who said the tragedy was investigated as a possible crime. When I called him at his home in Virginia, he said someone might have deliberately replaced the instructor's inert grenade with a live one—the scenario Aunt Bert had envisioned.

The instructor, whose name Canady said he couldn't recall, was a sergeant in his mid-twenties. He had used the same grenade nearly every day for weeks, keeping it in a metal container. His routine, Canady said, was a "get-your-attention type thing." He'd pull the safety pin and roll the grenade at the men to test their reaction, the idea being that you always had to be on the alert. But this time, Canady said, he might have picked up a live grenade that someone had planted. He might have thought it was the grenade he had always used. "When they had the training, there was supposed to be a practice grenade that was locked up in a cabinet and painted blue. Blue means safe." The grenade would not have had a fuse and a blasting cap, which acts as a detonator.

Canady, contradicting others, said MPs and the Americal Division's Criminal Investigation Division unit did investigate, "with a pipeline higher up." He had gone to the site and seen the damage to the orientation building. "I remember seeing the criminal investigators in their civilian suits. I was questioned about the training process— where training materials were kept." But, he said, the MPs found nothing suspicious and the CID investigation proved inconclusive. "They never determined who did it. It was never explained to me or anyone I know of."

Later, I asked Tony Viall and Tom Sled if they had been interviewed as part of an investigation. Both said they did not remember being questioned.

On July 23, 1969, Canady signed a Statement of Medical Examination and Duty Status that says under Details of Accident: "During period of classroom instruction, instructor unknowingly discharged a live grenade at approximately 1015 hours 10 July 1969, LZ Bayonet, RVN." The explosion was officially classified as an accident—and a former lieutenant colonel who had worked in the USARV Provost Marshal Office had told me that without verification of a criminal act, it's doubtful that a Report of Investigation would have been completed.

The instructor survived, Canady said, but suffered a mental breakdown. "He went bonkers, and I can understand why: You don't want to be responsible for killing your own guys. I don't know what finally became of him. I don't know if he was sent back to the States, discharged or put under [Veterans Administration] care." But who might have switched his grenades and why? If it had been a disgruntled American trying to frag the instructor, he would have known the explosion could also kill or maim others in the room. If it was sabotage by an enemy sapper, a specialist in the use of explosives who slips through defenses, how did he arrange it undetected, in daylight, in the middle of a busy LZ? Canady didn't know and wouldn't guess. He suggested I talk to the captain in charge of the Combat Center's instructors and gave me his name—Thomas C. Kerns, another Virginian.

Now I could barely contain my excitement. How could I get any closer than this? The man who led the instructors was sure to have answers.

To my amazement, he said he didn't.

When I spoke with Kerns in 1997, he said he didn't remember the explosion. "Logic says I was there" at Chu Lai when it happened, he said, "and I don't recall a thing."

A West Point graduate, Kerns had served with the 82nd Airborne Division in the Dominican Republic in 1965 and was in his second tour in Vietnam. In the first, he advised the South Vietnamese 7th Airborne Battalion, but that was cut short when he was shot in the hand and leg in a battle with a North Vietnamese unit. As operations officer of the American Combat Center at Chu Lai, he was responsible for all training activities. He said he supervised dozens of instructors, most of whom were senior noncommissioned officers who had teaching experience.

The idea of the Combat Center was to get new American troops acclimated to Vietnam. A week of courses covered survival, civil affairs and weapons training. There was the "scare class" on mines and booby traps that was held on a sand dune with the help of enemy soldiers who were prisoners.

Kerns said there were seven groups of new arrivals every week, each consisting of 120 men. Officers took a three-day course, while enlisted personnel had seven days of classes. Every morning, Kerns escorted the assistant division commander, a brigadier general, to a classroom where the general welcomed the new arrivals and talked about security, awareness, being part of a team and pride. But, Kerns remembered, the class where Nicky was mortally wounded was not part of the Combat Center course of instruction, though records I found later indicated it was. The instructors Kerns oversaw on the Chu Lai base, he said, did not use grenades, live or inert. "My instructors did not do that. No weapons were ever fired at the Combat Center."

In 2014, I re-connected with him by phone with the records I'd found in hopes that I could jog his memory. But he insisted that the Combat Center, which he proudly pointed to as including four "wonderful" classrooms, never sent anyone for training to LZ Bayonet.

A 1969 Combat Center schedule for replacements shows that the grenade class at Bayonet was part of the course, but the sergeant who taught it wasn't one of Kerns' instructors, just as Kerns had said. The schedule lists "movement to LZ Bayonet" and orientation there involving hand grenades as well as M60 machine guns, claymore mines, M79 grenade launchers and M72 light anti-tank weapons. The notation under "instructor" is NCOIC RNG, which stands for "noncommissioned officer in charge of the range." Missing was his name—the only one not on the schedule.

Kerns said he ran the Combat Center for a year and never saw a training schedule. There was no need for one, he said, because the classes were always held at the same times and places, with the same instructors, and notes sufficed.

"I do not recall a grenade explosion during a training class at LZ

Bayonet," he repeated to me in 2015. "Having never been to LZ Bayonet, I do not recall any training incident involving my instructors. I also do not recall there being any training [at LZ Bayonet] as part of the Combat Center's training program."

So, after all this unnerving back-and-forth, if Kerns wasn't in charge of the grenade instructor, who was?

I obtained Stanley L. Bartlett's name from Morning Reports for the week of the explosion. A captain, he was the 16th Combat Aviation Group's headquarters company commander.

Bartlett, who lived in Oregon, did remember the blast and said word of it was "all over" Chu Lai, as I would have thought would be the case. He said he had heard a rumor that a Vietnamese helper had laid out all of the instructor's props in the classroom that morning, then disappeared and was never seen again. A newsletter distributed by division headquarters to the headquarters companies across the base noted the incident, he said, and the aviation group commander mentioned it at a safety meeting. Beyond that, Bartlett could say only that he had gone through the same instruction at LZ Bayonet when he arrived at Chu Lai late in 1968. He was in the front row of the classroom when the instructor tossed a grenade and yelled, "Look out!"

General Ramsey had said he thought he had been briefed by someone from the 198th Light Infantry Brigade at LZ Bayonet—another possible lead. So I called the brigade's former commander, Jere O. Whittington of Texas, and was taken aback by his response, though at this point I probably shouldn't have been. The ex-colonel, who had led 5,000 men in three infantry battalions, an artillery battalion and a company of engineers, told me he had no knowledge of the explosion or an investigation.

I turned to Schwarzkopf. His autobiography had caught my attention for reasons other than his criticism of an aloof, out-of-touch Americal Division headquarters. He wrote that he had arrived in Vietnam in July 1969 for his second tour and worked at USARV, the Army headquarters at Long Binh, where he saw "plans and reports on everything from battles and casualties to budgets and USO shows."

The famous general did in fact answer me. But he too said he could not tell me anything. "I am sure an incident of the nature that you described would have been reported to USARV headquarters immediately," he said in a 1998 letter from his home in Florida. But he had arrived at Long Binh just after the classroom tragedy and did not remember seeing any report on it. He expressed disgust at Nicky's fate: "I am truly sorry that your cousin and your family had to be put through such a senseless tragedy. There are no excuses for what happened to him."

Colin Powell was another who might be able to help me. A major at the time, and on his second tour of duty in Vietnam, he was there as Ramsey's deputy assistant chief of staff for operations and planning. I wrote to him in 2005, after he had left the Bush administration as secretary of state. His response: "I was finishing my tour at that time, but assume I was still at division headquarters at Chu Lai. I have no recollection of the incident, and if any investigation was conducted, it most likely was completed after I left" on July 20.

Yet another key soldier at division headquarters was Lieutenant Colonel Francis A. Nerone, who had become assistant chief of staff for intelligence several days before the explosion. He is listed on the daily staff journal entry of July 10 as someone who was to be notified about what happened in the training class.

Once more, I struck out.

"I'm sure I was made aware of the tragic incident which resulted in your cousin's death," he wrote to me in 2013 from his home in Virginia. "However, responsibility for investigating it would have been outside my responsibility." He would have been involved, he said, if Nicky had died as a result of contact with the enemy.

Nerone did have a suggestion: He believed an investigation would have been handled by the commander of Americal Support Command, the umbrella unit for the replacement company, the Combat Center and several other units. He didn't remember who it was, but I found out: Colonel Joseph G. Clemons, Jr., a West Pointer and Korean War veteran who was portrayed by Gregory Peck in the 1959 film *Pork Chop Hill*. Clemons' first assignment when he arrived in Vietnam in July 1969 was to head Americal Support Command.

"That doesn't ring any bells," the eighty-five-year-old Clemons said in 2013 when I called him at his North Carolina home and asked about the explosion. "I really don't remember that."

This was just one more of the many disappointments that piled up in my contact with commanders, chief among them Ramsey, who had been charming and friendly but offered little help. He had been the ultimate authority at the Chu Lai base—the man to whom all vital information concerning the Americal Division flowed—but he could not remember key details surrounding a tragedy that happened under his command, a singular, extraordinary event that he said had shocked him for its horror and senseless waste of soldiers' lives.

He had remembered the particulars of his own adventures and relished passing them on. That fall of 1998 when Mary and I visited him, he told stories connecting him to some of the great figures and events of World War II. As an aide to the British general who was the ground commander under General Dwight D. Eisenhower in North Africa, he said, he had once sat with Prime Minister Winston Churchill at a dinner. "Where are you from?" Churchill asked him. "I'm from Kentucky, sir, where the horses are beautiful, the women are fast and the liquor is fine." As an officer in the 3rd Infantry Division, he had met legendary correspondent Ernie Pyle and knew Audie Murphy, the most decorated American soldier of the war, "a farm boy who wasn't afraid of anything and wasn't afraid to do anything," he said. Then in the last days of the war in Europe, twenty-six-year-old Lieutenant Colonel Ramsey stood near Berchtesgaden, Germany, site of Adolf Hitler's Alpine headquarters, for the raising of the American flag.

One of the letters Uncle Louie received in the weeks after Nicky's death came from Ramsey. The general opined about "this vast and most cruel of wars" and said Nicky "was a most courageous officer and his actions here in Vietnam were in the highest traditions of the United States Army. We shall all be poorer for his loss." It was a form letter, of course. Ramsey didn't know Nicky, and other than attending Combat Center classes, my cousin hadn't done anything at all. He hadn't even set foot inside a helicopter.

When I asked Ramsey what he thought about the instructor's grenade routine, he echoed Bacon: It wasn't proper, and if he'd known it was going on in the classroom, he would have stopped it.

But a former MP who remembered the explosion told me that soldiers who had been in Vietnam for a while played all kinds of tricks on new arrivals to embarrass or frighten them.

Ramsey, who said he had many sad memories from Vietnam, once wrote to me, "War is hell, but it's even worse when you lose your life in an accident like [Warrant Officer] Venditti did."

I just wanted to know the details of what happened and if there was any follow-up. Why couldn't those in charge tell me?

It was time to move on resolutely in my pursuit.

Many Vietnam veterans were out there with stories to tell—the grunts of lower rank. If I wasn't going to get answers from Chu Lai's commanders, I would see what the men under them had to say.

Chapter 12

Patchwork of Memories, 1996–2015

As the '90s ticked away and the millennium approached, I had to reconcile myself to the fact that no set of documents, none of the men in charge, no single individual was going to hand me the full picture of what happened that day at Chu Lai. So, over the course of years, I slowly constructed a mosaic, bits and pieces from the sad recollections of dozens of men and women from all over the country—a small detail here, a fleeting memory there. In place of dramatic revelation, my quest evolved into slogging inquiry: searching numerous veterans' Internet sites and publications, mining Nicky's files for names, matching them with addresses and phone numbers, making calls, writing letters.

A steady drumbeat of emails, snail mail and phone conversations followed. Some accounts were first-hand, while others came from vets who were remembering what someone told them or what they over-heard. All were recalled through a filter of thirty years' time, a challenge even to people with the sharpest memories and the keenest eye for details. Accounts conflicted, and some seemed wildly off the mark. Some of the vets strongly believed there had been a cover-up of the incident and were still angry. Some lashed out at the stupidity of the instructor's toss-the-grenade routine for the great danger it carried. Still others saw the explosion as a dark act of sabotage by the enemy.

All along I hoped for the improbable—that I would hear from the instructor himself, eager to tell his story, something he perhaps had kept to himself for so long and now had an opportunity to reveal. Short of that, I wanted to talk with anyone who had been in the classroom when the grenade went off. Thirty or forty men had been there. The two I had spoken with—Tony Viall and Tom Sled—had been badly hurt.

Most of the others had minor wounds and presumably went on with their tours of duty. Who were they and where were they?

One day in late December 1997, a tantalizing one-line message showed up in my email. I caught my breath and felt a bolt of anticipation. All it said was: I WAS THERE JULY 10, 1969. Below were a name, Sam Pilkinton, and a phone number.

I wasted no time in calling him. He was a catfish farmer in Mississippi who had seen my query in the Americal Division newsletter. He told me a story he said he hadn't thought about in twenty-five years. He said he had been in the classroom at LZ Bayonet, had recognized that the grenade was live the moment it left the instructor's hand and had saved himself in a fleeting act of desperation.

Pilkinton told me that he arrived in Vietnam on July 4, the same day Nicky had. The nineteen-year-old combat engineer, who would later be assigned to the 26th Engineer Battalion, had more experience handling grenades than most other soldiers. After advanced training at Fort Leonard Wood in Missouri, he had been held over for several months and worked at its grenade range, where he threw several live ones and a few hundred practice grenades, which worked just like the real thing.

In the orientation building the day of the explosion, Pilkinton said, he sat on a bench behind and to the right of where the grenade went off. The instructor had a wooden box containing about two dozen high-explosive M26 grenades, and though Private Pilkinton couldn't tell if it was full, he said, he could see the tops were off some of the cardboard holders. He described the instructor as cocky. "He had been there a while and was trying to say, 'You better pay attention or you won't live as long as I have.'"

The instructor picked a grenade out of the box. Pilkinton felt uneasy but knew the sergeant didn't intend a demonstration with a live grenade and assumed it had been disarmed before the class began. A fragmentation grenade has a pull ring and safety pin. Removing them allows a safety lever to fly off, he knew, but if you hold the lever in place with your thumb, nothing will happen. When you throw the grenade, the lever comes off by the action of a striker and spring. The striker rotates and hits a primer, which emits an intense flash of heat. That flash activates a time-delay element that burns for about five seconds down to a detonator, which is like a blasting cap. It explodes the grenade.

To disarm the M26, Pilkinton said, the instructor would have "shot" the fuse. That meant he'd unscrewed the fuse—an assembly that included the safety lever and pull ring—from the body of the grenade, pulled the pin on it, tossed it aside and let it detonate harmlessly. Then he would have picked up the spent fuse, screwed it back into the M26 and returned the grenade to the box, where it would appear to be real and newly shipped, more likely to frighten new arrivals than a practice grenade clearly marked in blue.

Pilkinton said he watched intently as the instructor pulled the safety pin and flung the grenade. When the safety lever flew off, he said, he heard a *pop*—the sound of the striker hitting the primer, setting fire to the fuse. No such sound would have emanated from a disarmed grenade. "There was fear in my heart," Pilkinton said, "and I got right outside. It was my instant, gut reaction. I knew the grenade was going to go off in five seconds."

The explosion and pandemonium followed. Later, he remembered, there was talk among soldiers outside the building. They had heard that this instructor used disarmed grenades for his class instead of the M30 practice grenade. The M30 was painted light blue and was the same size and weight as the M26, but its fuse emitted a puff of smoke and there was a sharp report similar to a firecracker. Using an M26 seemed to Pilkinton like playing with fire.

Immediately afterward, Pilkinton remembered, he offered to explain the circumstances to investigators but was told to shut up. The next day he was sent to LZ Fat City, west of Chu Lai, where investigators came to talk to him "but didn't want to hear what I had to say."

"This was a stupid accident caused by untrained, unsupervised soldiers playing stupid games," said Pilkinton, who went on to serve in the Persian Gulf War and retired from the Army as a chief warrant officer four in 1993. He felt that "the leadership should have been held responsible for what happened. The captains and colonels should have been accountable."

Army records confirm that Pilkinton arrived in Vietnam on July 4, as he'd told me, and that he was with the American Division at Chu Lai when the explosion happened. But I could not independently confirm his account.

One of my other discoveries shed light on how the grenade lecture was meant to play out. It also offered an intriguing new detail that would crop up elsewhere.

The source was Steve Stone of Iowa, a helicopter pilot who had been in Nicky's orientation group at the American Combat Center. They were among 100 men divided into two groups for POR instruction—Process for Overseas Replacement—as they were being moved into the Americal Division. Stone attended the class at LZ Bayonet the day before Nicky and his friends. He said the instructor strode back and forth on the classroom stage, gripping a grenade in front of about forty soldiers. On a table up front were several kinds of grenades and some other explosives. A few other instructors stood in the back.

"Uh-oh," the instructor said as he pulled the pin and dropped the grenade in a half-toss, as if he didn't mean to do it. He went behind a lectern and stepped on a switch. The grenade rolled to a stop in front of Stone. Blindly trusting that it wasn't a live grenade, and knowing that the instructor was making the point that they were in a danger zone, Stone got up off his bench and bent down to pick up the grenade. Just then, a charge activated by the switch went off outside the building, a muffled *whump* maybe thirty yards away.

As Stone reached for the grenade, which had a blue stripe indicating it was safe, he considered what to do with it. He didn't want to play the instructor's panic game, so he thought about throwing it out the door. But he decided against that, uncertain how the instructor would react—he didn't want to give the impression he was clowning around. He clutched the grenade and underhanded it to the instructor, who caught it.

The next morning, while Nicky was trucked out to LZ Bayonet, Stone was back at the Combat Center going through the "gas chamber," the building filled with tear gas that the new arrivals had to navigate without wearing a mask. Afterward, when he went to the barracks to shower, he learned about the explosion from another helicopter pilot, Larry Feasel, who should have been in the grenade class with Nicky but had skipped it. Stone immediately remembered the grenade that had been at his feet the day before and thought: That could have been me! "I remember Larry commenting that had he not skipped out that day, he too could have been one of the casualties from the incident. And we decided to go get something stronger to drink than just a soda."

Stone went on to receive a Bronze Star for service and a Distinguished Flying Cross for a medevac mission during his time in Vietnam, and he would fly Hueys again in the Persian Gulf War as a medevac pilot. His story about the grenade lecture introduced a new detail in the classroom scenario: that the instructor stepped on a switch to set off a charge outside the building as the grenade rolled, apparently to give the new guys a fright.

Feasel, who lives in Indiana and saw my query in the Vietnam Helicopter Pilots Association newsletter, said he had skipped the grenade lecture because he was fed up with the training. Instead, he and another warrant officer went to the PX, a store that sells equipment and provisions. He gambled that he wouldn't get caught for cutting the class. Around 3 p.m., a portly captain whose name he didn't remember showed up at the barracks. "He was pale white and very upset, almost on the verge of breakdown. He said there had been an accident in the classroom and explained what had happened." The captain had been in the class and had flecks of blood on his face and clothes, Feasel said.

The instructor, a veteran infantry noncommissioned officer, was explaining hand grenades, Feasel said the captain related. He had a box of grenades on a table. He took one of the round cardboard holders with a grenade in it and pulled off the tape securing the two halves together, as you would with a new out-of-the-box grenade. He held the grenade up and said: "This is an M26 grenade." He pulled the safety pin and held the lever to show the final safety, and then he let the lever pop off and threw the grenade toward the class. It landed on the floor and rolled under the bench where some warrant officers were sitting. One of them kicked the grenade aside, and it rolled under Nicky's bench. The instructor stepped on a button on the floor, setting off a charge outside the building, according to the captain. "There was a loud popping sound," Feasel wrote, "and then the second explosion of the live grenade going off inside the classroom."

Other veterans discounted the idea of an accident and pointed instead to a shrewd enemy at work behind the scenes. They echoed what the replacement company commander had told me—that the explosion might have happened according to plan, set in motion by someone who made sure the instructor had a live grenade when he walked into the classroom. Such accounts added to the confusing jumble I had to sort through.

A grunt from Texas who served in the 501st Infantry, 101st Airborne Division was wounded July 18, 1969, at Tam Ky, just up the coast from Chu Lai. He emailed me that he was evacuated to the U.S. Army hospital at Camp Oji, Japan, and that the soldier in the bed next to him was from the Americal Division. He thought his name was Jim.

Jim said he reported to Chu Lai for training immediately after arriving in-country. One day an instructor walked into a room full of new troops, picked up a grenade and threw it toward them. It detonated and Jim was hit by fragments. He said he'd been told that an enemy agent was responsible—Vietnamese working on the bases included some enemy personnel. Jim said he'd heard that a civilian worker had swapped the instructor's inert demonstration grenade with a live one, knowing it would be tossed in the class. The enemy agent had been found and dealt with, Jim said, and the instructor was one of the dead.

But the instructor did not die, according to the Army casualty records I had pored over and what Doug Howard of the Army's Casualty and Memorial Affairs Operations Center had told me. And I had found no records or other information about a Vietnamese worker's involvement or of a criminal investigation that pinned the explosion on sabotage. Moreover, the Americal Division's Operational Report/ Lessons Learned for that period noted there had been no incidents of sabotage.

A few soldiers I reached had been outside the orientation building when the grenade went off and described the immediate aftermath. Tad Eversole of the 1st Battalion, 14th Artillery had spent the night of July 9 on guard duty in one of the observation posts along LZ Bayonet's perimeter. About 10:15 the next morning, he was walking toward his unit's mess hall, carrying his M16 rifle and a bandolier of ammunition. The orientation building stood directly in his line of sight. When he was fifty yards away, he heard a loud, muffled explosion and saw a puff of smoke. He crouched down for a moment, thinking it was an incoming Viet Cong mortar round—LZ Bayonet had been mortared in the past, though only at night. Then he and others charged toward the building.

Eversole, of Washington State, said that later in the day, he and about ten others drinking beer and sodas in the enlisted men's club at LZ Bayonet wondered: How could this happen? One of the instructors who had been observing in the classroom came into the club and said there were five or six dead. He said that when the sergeant flipped the grenade, the observers realized—and the sergeant did, too—that it was live. They heard a *pop* and saw a puff of smoke that meant the striker had hit the primer, lighting the fuse. All of the instructors froze. Eversole heard the sergeant was busted—he lost his rank—and was sent out to the field.

Eversole's statement that the instructor recognized that his grenade was live when it left his hand was a story I heard more than once.

On a helicopter pad seventy-five yards from the orientation building, Stanley Elliott of the 1st Battalion, 52nd Infantry was about to board a chopper bound for a forward firebase, LZ Stinson, to deliver supplies to his company. He heard the unmistakable *thump* of ordnance going off and felt the whiz of metal flying over his head. He turned and saw men running out of the classroom, then medics from his battalion headquarters getting there almost immediately. One medic fell off the back of a jeep ambulance that hit a bump as it sped toward the building. The jeep stopped and he jumped back on, said Elliott, of Oregon.

Don Stuhr of the 1st Battalion, 14th Artillery heard yelling and screaming and saw men charging out of the building as if it were on fire. He and a buddy rushed to help and came on a scene of chaos, with blood all over, torn men, faces in shock and exclamations of "What happened?" Stuhr helped get men out of the building and load the wounded onto helicopters. He assumed that four to six soldiers were fatally injured. That afternoon, he and nine other men washed the blood away from the classroom, using water trucks. Gobs of blood stained walls, tables, benches and the floor, and grenade fragments were embedded in the tables, he remembered.

Stuhr, an Iowa farmer, said he had attended the grenade class during orientation when he came to Chu Lai in March 1969. The instructor pulled the pin, threw the grenade down the center aisle and said, "You've got five seconds, what are you going to do?" The new guys scrambled to hit the floor.

David Holdridge of Louisiana called me one day to say that he had driven one of the two deuce-and-a-half trucks that took the new arrivals from the Combat Center to LZ Bayonet that day. He pulled up alongside the orientation building, stopped, got out and let the tailgate down. He said about thirty soldiers went inside, and two drill sergeants stood in the back of the classroom "to make sure things went smoothly."

He was on the scene immediately after the explosion and said it looked like "all hell.... They were pulling people out. We got two guys into the jeep ambulance. A couple other jeeps were there. I helped bandage one guy." Later he drove the remaining soldiers back to the base—sixteen to eighteen men who had been treated by medics and were not badly hurt. Holdridge said he thought the instructor was from California.

A former MP, who was assigned to the 198th Light Infantry Brigade at LZ Bayonet for a year starting in the fall of 1968, said a half-day at the LZ was part of in-country orientation for the "newbies."

"When I went through the Combat Center and spent my half-day at Bayonet, the orientation was outside at a firing range. We all sat in bleachers and the instructors were down in front where demonstrations were given on just about any weapon that we may use while in 'Nam. The grenade was thrown down-range, and then a dummy was tossed into the bleachers. I thought at the time, 'This is not funny.'"

On July 10, 1969, the MP was at Bayonet's front gate, about a quarter-mile from the orientation building and firing range, when the instructor's grenade went off. "As the range was 'live' at the time, I doubt anyone thought that anything had gone wrong until medevac units started to arrive from Chu Lai. All of the MPs from Bayonet who were on the base at the time were called to secure the area." But, he said, none of Bayonet's MPs had anything to do with an investigation. His group on the LZ, he said, was the 3rd Platoon of the 23rd MP Company. They were combat MPs who worked mostly on convoy escort and Highway 1 patrol. He said the 1st Platoon or the CID out of Chu Lai would have conducted an investigation.

"The incident was one of the worst experiences of my year there, in that I have always figured that it was something that never should have happened."

Soldiers across the base were shocked and saddened, especially those who were in the business of teaching newcomers how to fight and stay alive. "It was a bad day for all of us," said Clyde Curtis of Louisiana, a Combat Center instructor who specialized in mines and booby traps. He said he thought "the instructor used poor judgment."

Darryl James of Texas, a pilot and captain in the division's artillery aviation section, Division Artillery Air, had been at Chu Lai nearly a year when the blast happened. "News swept through the division like wildfire," he said. "We all wondered how such a tragedy could happen." And he noted that lectures that required the attendance of newly arrived pilots had not been a longstanding practice. "When I came to Chu Lai in September 1968, aviators were not put through that orientation course that your cousin was in. We went directly to our assigned unit."

Edward Geserick, Jr., heard about the grenade's going off and the psychological toll it exacted on the instructor. Geserick had been a cook at the Combat Center but was reassigned in June 1969 to visual aids because he was an artist. He created training posters and illustrations to promote Vietnamization, the program to turn the fighting over to the South Vietnamese. "Scuttlebutt was that it was customary for the instructor to pull this stunt, to throw the grenade at the crowd," said Geserick, of Florida. "I heard one person died and the instructor really went crazy. He had a complete mental breakdown. He was terribly upset, full of guilty feelings. I heard he went to a hospital for treatment."

Bob Short, who in June 1969 joined the 46th Infantry, said that "after my year in the infantry, I stayed in Vietnam for six more months as an instructor at the American Division's Combat Center. The incident in which your cousin was killed was still being used as an example of how not to conduct classes with new arrivals."

Short, of Michigan, became an instructor in the spring of 1970, about ten months after the tragedy. His statement, delivered from memory, was the only time I heard that the grenade routine was halted after Nicky, Billy Vachon and Tim Williams died.

A comment posted on my blog site in the spring of 2013 raised my hopes to a high pitch. I had mentioned Nicky and what happened to him. A Chu Lai veteran in Minnesota, Ralph Anderson, saw it and wrote that a buddy of his in the 723rd Maintenance Company had been in the classroom and survived the explosion without a scratch. When I followed up with Anderson, he was sure that Bob Beck could give me a lucid account because he had been sitting in the back and had seen everything that transpired.

"I remember well the day Bob showed up at the 723rd," Anderson wrote. "He was a nervous wreck and still almost deaf from that grenade, and all he wanted to do was stay as drunk as possible for as long as possible. After about three or four days, he sobered up and told us all about it."

Anderson said Beck didn't know that anyone had been killed. He only mentioned that some guys lost their feet and some lost their hands. Here's what Anderson remembered of Beck's story after forty-four years: The instructor, a sergeant, pulled the pin on a grenade, tossed the grenade in front of the guys in the first row and hollered at one of them, "Throw it out the door!" The guy jumped up from his bench and ran up to the grenade but didn't pick it up. Instead, he ran out so fast that he actually tore the screen door off the building. The next second, *wham*!

Where could I find Beck? Anderson remembered that he was from Doylestown, Pennsylvania, right near Allentown—I used to work at the Doylestown newspaper. But it's where Beck lived when he went into the Army, and he and Anderson had no contact after Vietnam. Where was he now?

Bob Beck, who might have given me a sharper picture of what happened, died in 2010 of cancer at age sixty-three. He had never left eastern Pennsylvania. At the time of his death, his home was an hour from my own.

Some of the veterans I found had been hospital staffers at Chu Lai. Dr. Alton Gross of Virginia, who operated on Nicky at the 27th Surgical Hospital, remembered that about thirty casualties were dispatched to his hospital and the 312th Evacuation Hospital on July 10, 1969. Of those, a half-dozen had severe injuries.

News of the explosion sickened Dr. Bradley Billington of Washington State. He saw little point in all of the new guys sitting through the grenade lecture together, but was told the idea that everyone should go through the same training fostered *esprit de corps*. When the instructor tossed a grenade during the class Billington attended, he thought it was a stupid and dangerous trick and says he confronted the instructor angrily. "Take it easy, Doc," the sergeant told him. "It's just a way to get their attention." Months later, Billington helped Gross amputate Nicky's leg at the 27th Surgical.

Operating room technician Dave Knox of Alaska was on duty at that hospital when Nicky and some of the other casualties came in from the LZ. He had been in that same class about two weeks earlier and the instructor tossed a grenade into his group. "Even though this grenade was a dummy, it was a very scary experience," he said. "No one had time or the sense to do more than hit the deck. We were all so rattled after that, I think they gave us a break to go outside so that the guys who smoked could light up."

Some focused on the high quality of medical care soldiers received. I found that veterans were eager to talk to me about it, that they felt a need to share memories of people and events that had affected them deeply. One of them was Larry Baker of Illinois, who had been a patient in the same ward as Nicky at the 312th Evac and remembered hearing nurses talk about what happened. Baker, of the 17th Cavalry in the 198th Light Infantry Brigade, was wounded three times during 1968 and '69.

"The last time I was wounded was the Fourth of July 1969, by a land mine explosion. My wounds were so severe that I had to stay in-country until the 18th of July before they flew me to Japan. I lost my left leg above the knee and had many chest and lung wounds.... I knew without a doubt my life was saved at the hospital, and I am sure that Nicky received the best treatment possible."

Troops in the field knew that if they got hurt, they would get the best of care. Short, who was a squad leader with Charlie Company, 1st

Battalion, 46th Infantry before he became an instructor at the Combat Center, attested to that: "One of the things that kept those of us in the bush going day to day was knowing that there were doctors, nurses, helicopter pilots and other support personnel who would do anything humanly possible to save us if we were seriously injured."

That confidence was borne out by the statistics. Retired Brigadier General Anna Mae McCabe Hays, who had grown up in Allentown and led the Army Nurse Corps at the peak of the Vietnam War, said in an Army interview that a wounded U.S. soldier was likely to survive, thanks to combat medics and the staff of field hospitals. Only 1.2 percent of wounded Americans who lived long enough to reach a hospital in Vietnam died after getting there. That was about 2,500 men, a fact that only highlights the tragedy of Nicky and Billy.

Despite all the grief and horror, many got through their time in Vietnam and made it home with minds and bodies intact. Larry Feasel and Steve Stone, the two helicopter pilots who just missed being with Nicky in the ill-fated classroom, would spend a year with the 176th Assault Helicopter Company, the proud unit that Nicky and his friends had hoped to join. It had about sixty pilots and the same number of crewmembers. Stone said fewer than ten men were lost.

It was clear what that meant. If Nicky and Billy had only gotten past their week of orientation, they probably would have survived fifty-one more weeks. Instead, they and Tim Williams were among the sixteen percent of U.S. military deaths in Vietnam that the Pentagon attributed to accidents.

There was no point in dwelling on the small odds that Nicky would be in an accident and that a well-appointed hospital with an outstanding staff would not be able to save him—a one-two punch of unlikely outcomes. Nicky, Billy and Tim had walked into that orientation building on the one morning that something went terribly wrong. And when the grenade rolled on its haphazard path, they were in seats that put them exactly in line for its hot metal fragments to catch them and do the most damage.

But I could not lie back and weigh the what-ifs. I had to move on.

The push would soon lead me to the identity and whereabouts of the man whose name appeared nowhere in the scant records—the instructor. But that would not come easy.

Chapter 13

No Longer a Phantom, April 30, 2000

"I'm not going to tell you his name."

The tone was defiant, tense. The man was determined not to give up what I needed. I felt panic as our clipped phone conversation seemingly moved toward ending. I couldn't let that happen. Just keep him talking, I thought, and maybe he would lower his guard. I had to connect with him, gain his confidence, draw him out.

I had to because this former Army medic, a teacher I will identify only as "Doc," a common tag for combat medics, was the linchpin.

For almost six years, the huge missing piece in my search for the details about Nicky's death was the identity of the instructor who had hurled the fatal grenade. He was a phantom, excruciatingly elusive. Everywhere I'd turned to learn his name, I had come up empty-handed. It was not in any documentation in the National Archives or any Army holdings, not in Nicky's records or those of Billy Vachon and Tim Williams, not in the memories of the commanders at Chu Lai, not even on an orientation schedule that listed every instructor except the one teaching the grenade class in July of 1969.

But here I was at last, on the phone with the man who knew.

Yes, he'd gotten my letter, he had said moments earlier from his home after answering my call. Then right away he demanded to know what I understood about the practice of throwing dummy grenades in orientation classes.

I said Americal Division veterans had told me they'd attended classes at Chu Lai where the instructor tossed an inert grenade at them.

That's right, he said, it was a common practice done for an important reason: to train guys how to react. Good instructors taught men to expect the unexpected, always to be on their guard. That was how they'd survive their tours of duty.

"Who threw the grenade that turned out to be live?" I asked.

Doc bristled, refusing to say.

Finding him had been the result of a slow but deliberate process beginning with a story I had written about Nicky that ran in *The Morning Call* in September 1999. I sent a copy to the Vietnam editor of the Americal Division Veterans Association newsletter, and he invited me to write an article for the publication. Thousands of the unit's veterans got that newsletter in the mail. I hoped it would shake someone loose.

The newsletter story ran with photos of Nicky, Billy and Tim, and some from my 1998 trip to Vietnam, and included my phone number, address and email. Weeks later, the editor wrote to me that he had gotten a call from a veteran in Illinois who had read the article and said he knew the sergeant who pulled the pin on the grenade. I put the letter down, picked it up and read that sentence again. Then I read it a third time, slowly, aloud.

"I think he was going to contact you, but he may have reservations," the editor wrote of Doc, who was then fifty-five years old. "You must understand that it is sometimes difficult for some to contact family members of those lost in Vietnam." He included the man's address.

This was the break I needed. Without any official Army documents, my only hope lay in finding soldiers like Doc who'd been with the Americal Division in the summer of '69. I wrote to him that I had no ill will toward the instructor, only a desire for answers. I closed by saying I'd call him in a week.

For the next several days, it was hard to think of anything else. Would my letter work? What next if he refused to tell me the instructor's name?

Sunday, April 30, 2000, was the day I'd promised Doc that I'd call. It was the twenty-fifth anniversary of the end of the war, the day a Vietnam commemorative section I'd edited appeared in *The Morning Call*. That night, I fought the fluttering in my chest and, my hands shaking, picked up the phone and pressed the digits of his number.

After a couple of rings, he answered.

To my relief, he did not cut me off after curtly refusing to tell me the instructor's name, and I didn't need to prompt him to keep talking. Instead he offered to relate the circumstances surrounding that soldier's misfortune. Like other veterans who had seen courage and suffering, he was eager to tell me something of the experience of war. His comment on the use of grenades for instruction showed he wanted me to see the instructor not as a demon to be vilified, but as a casualty to be pitied.

That's because he had known the instructor well. They were about as close as two infantrymen could get in Vietnam.

Doc was drafted while teaching at a high school in Chicago, arrived in Vietnam as an Army private in the spring of 1969 and was assigned to the 46th Infantry, then part of the 198th Light Infantry Brigade. At twenty-three, the college graduate was one of the older guys.

During his orientation at the Americal Combat Center, instructors in the mine class "told you how many times you killed yourself or others." In the grenade class, the instructor whispered to one new arrival to jump out the window when he tossed a grenade, and whispered to Doc to rush out the door. He and the other GI did as they were told. No one else in the class tried to get out. The instructor said the two who bolted would have been the only survivors if the grenade had been live.

Doc met members of his unit while they were on stand-down at Chu Lai. He was one of five company replacements. A sergeant introduced him around, and he got the job of assistant on an M60 machine gun, which meant he would be Hesston's assistant, he said.

A bolt swept through me and my grip on the phone tightened.

Who was Hesston? I interrupted in as casual a tone as I could muster. I didn't want him to mistake the question for a demand.

There was a long pause, silence on the other end of the line. A rising, urgent tension crackled. I knew immediately what that meant but tried to stay calm until he spoke again, hoped he wouldn't hang up.

"That's him. He was the instructor."

I heard myself gamble on getting more: "What was his first name?"

"Wayne. Wayne Hesston.* He was my mentor."

I caught myself from an involuntary sucking-in of breath. After all the years of searching, there it was—a simple name, an everyman's name. I turned it over and over in my mind in a flash, wanted to shout that the man who killed Nicky wasn't a phantom anymore.

Doc had done what I hoped he would do—let slip the name. Now that he had given me what he swore he wouldn't, he was not about to retreat. He wanted to keep talking, determined to impress on me that this man had been no ordinary grunt, but a soldier to be admired.

*Wayne Hesston is a pseudonym. See Chapter 18.

Hesston had been with the 46th Infantry for five months by spring 1969, Doc said, and was tough and resourceful, lean and strong. He was three years younger than Doc but had a wealth of experience. "If it wasn't for him, I wouldn't have learned so quickly. He taught me how to survive in the jungle."

The two saw a lot of action during the twenty-nine days they were a machine-gun team, and Hesston was brave. During one enemy attack, he and his assistant were hunched together with the M60 and its ammo. Intense incoming fire scattered and unnerved the men. Doc heard his teammate gag in pain and turned to see a large splinter sticking out of Hesston's throat. Hesston yanked it out, got up with the machine gun and, without a helmet on, ran across the clearing in front of them to the other side of the perimeter, firing and rallying his buddies.

In the spring of '69, Doc said, Hesston got his sergeant's stripes and gave up "the pig"—Army slang for the eighteen-pound M60. One of Hesston's knees was going bad, and on a trip to Chu Lai, he'd made it known he was interested in becoming an instructor there. He had the right qualifications: He was a sergeant and had experience in the field. In June, he went to the rear, to Chu Lai, and began to teach new arrivals.

The picture Doc painted of Hesston as a commendable infantry-man, a good man in the field, conflicted with the image I'd had of him from conjecture and the bits and pieces in the stories I'd picked up over the years. I hadn't imagined him as a soldier anyone would be proud to serve alongside, but rather as arrogant, a showoff—and thus likely to be the kind of sergeant who would use a grenade to play a dirty trick on a bunch of replacements just arrived in the country.

When Hesston took on his new job as an instructor, his assistant stayed in the field and became a medic, earning the nickname Doc. By July it had been a while since he'd seen his friend and mentor.

Then one day, a grunt who had been to the rear came back to the landing zone, LZ Professional, with shocking news: Hesston had thrown a grenade while teaching a class and it had detonated. A few men died, and many were hurt. Doc was shocked and felt terrible for his friend, so he went to Chu Lai within a couple of weeks determined to comfort him. "I felt close to him," he said, but Hesston refused to see him or anyone else. As Doc told me this, I felt a surge of empathy for Hes-ston—suddenly he was more human to me. He was a person tormented by what he had done.

Had Hesston been hurt in the explosion?

Doc didn't know. He had never spoken with or seen Hesston again.

What did he hear about what happened?

He remembered talk that a Vietnamese hooch boy was responsible. The story went that the boy had helped Hesston with his supplies, laying out the classroom's props early in the morning, before class, and also had worked for the previous instructor. But after the explosion, no one saw the boy again. It was an account that clashed with the Americal Division report of no incidents of sabotage during that period.

Years later, Doc and another veteran wrote to Hesston, inviting him to a unit reunion in the late 1980s. But Hesston didn't come and Doc didn't know whether he had even responded to the invitation. At the time, he said, Hesston was living in California and driving a tractor-trailer.

Doc wouldn't give me Hesston's last known address—he wasn't sure whether he had it, anyway.

I thanked him and we said goodbye. It was late, around eleven, and the house was dark. I wandered down the hall and slipped in and out of a succession of rooms in a daze, recounting what Doc had said, repeating the highlights. After years of searching, I had a name.

Wayne Hesston.

Apparently he was still alive, living in California, driving a truck. *Living in California, driving a truck.*

Mary was in the kitchen. I told her the news, and she shared my awe. We both recognized an added dimension to the story that enhanced the magnitude of the tragedy that summer day in the classroom at LZ Bayonet.

For years, the instructor whose grenade gimmick killed Nicky, Billy Vachon and Tim Williams had always been just that—the instructor. He was the bad guy, a smartass who pulled a stupid stunt, who failed to make sure his props were safe, who should have faced a charge such as negligent homicide, who should have suffered swift, harsh punishment for his recklessness. From the start of my search, when I knew only the outline of what had happened, it was hard to feel charitable toward the man who had wasted Nicky's life, even if it had been an accident.

But now here was Doc, who did more than give the instructor a name. In two hours on the phone, he had shown me a young man who had been a good soldier and a leader, who had faced death bravely, who had done his duty trying to teach replacements how to stay alive—only to experience the horror of causing some of them to die.

Now I had a name and a realistic hope that I might actually find the man.

At first I found a dozen listings for the man in California and couldn't narrow the search because I didn't have his middle initial. Then I checked a Vietnam database website that aimed to list the names of everyone who served in the war. That yielded four names matching the instructor's—three soldiers and an airman.

As I'd done before when I was stumped trying to find Vietnam vets, I turned to Dick Bielen of the U.S. Locator Service in St. Louis. His business was finding people and military records. I'd discovered him in 1999 after reading a series in the St. Paul (Minnesota) *Pioneer Press* about Vietnam vet Jon Hovde, an amputee and author of *Left for Dead: A Second Life After Vietnam*. The five-part series caught the attention of a newsroom librarian at *The Morning Call* who was interested in my project. He gave me copies of the St. Paul paper, which told Hovde's story and Bielen's role in reuniting him with a nurse who had tended to him after he was severely wounded. I emailed the reporter who wrote the series, and he put me in touch with Hovde and the people who had helped him, who in turn steered me to Bielen in St. Louis.

A Vietnam veteran and former journalist, Bielen had always come through for me. I had told him Nicky's story, and in the past year-and-a-half he had advised me on where to turn and how to get information from the military. He provided contacts, Army records, phone numbers and addresses. When I couldn't find a Vietnam vet I needed to reach, he could, among them Samir Marrash, the doctor who tended to Nicky at the 312th Evacuation Hospital.

I called Bielen and told him what I knew about Hesston. The next day, he called to say he had narrowed the possibilities to one soldier. His military occupational specialty code indicated he was an indirect fire infantryman, such as a mortar operator. He was born in 1948. He had been discharged from the Army in the early '70s, which jibed with the six-year obligation at the time.

Bielen gave me Hesston's street address, phone number, month and year of birth.

"What are you going to do now?" he asked.

"I'm not sure. Maybe I'll write Hesston a letter," I said.

"Don't do that yet. Let's make absolutely sure we have the right man."

"How do we do that?"

"Get copies of his military records."

Bielen faxed me a form he uses, and I requested Hesston's records from Army Personnel Command in St. Louis under the Freedom of Information Act. I didn't need Hesston's permission because a person's military service is a matter of public record. But under the Privacy Act, I wasn't entitled to personal information about him, such as his birth date and place of birth, the names of his dependents and his home of record.

Bielen had given me the name and number of a contact at Army Personnel Command, case analyst Peggy Barton of the Special Inquiries Team. After a month, I called her to make sure she had my request. She had it and was processing it.

Another month went by, and on July 13, 2000, I called again. This time she had Hesston's records at her desk. They still needed to go through a couple of staffers who would cross out information not releasable under FOIA, but she paged through them and read parts to me. And what she told me confirmed that Wayne Hesston was the man I sought. He had served in Vietnam with the 46th Infantry. So far, Doc's information was spot-on.

On August 5, almost three months after I faxed my request, a large manila envelope arrived from the Department of the Army, U.S. Army Reserve Personnel Command. It contained fifteen pages from Hesston's personnel file. The first thing I noticed: There was nothing about the explosion.

One paper was an order for Hesston to report to the Specialized Treatment Facility at Fort Ord, California, for an unstated period. The reporting date was almost two months after the grenade blast, but shortly before his tour of duty in Vietnam was supposed to end. Another document shows Hesston was awarded a Purple Heart for a wound he suffered early in 1969. He was assigned as an instructor at the Americal Combat Center shortly before the explosion. A personal history, which Hesston gave when he enlisted, put him at five feet ten inches tall, weighing 190 pounds. He had blue eyes and brown hair. He graduated from high school in 1966.

Hesston's application for his first Army ID card had a grainy picture of him the size of a postage stamp. It showed a man with close-cropped hair and a stern expression. He looked hardened and older than his nineteen years. I wondered if that's why some soldiers had told me the instructor was in his mid-twenties, when actually Hesston was just twenty or twenty-one at the time. Other papers indicated Hesston had trained at Fort Benning, Georgia, left for Vietnam in 1968, received his Combat Infantryman Badge and was promoted from private first class to specialist fourth class and then to sergeant.

In the weeks after I got Hesston's records, I spread the word that I had identified the instructor. I told Aunt Bert and Nicky's mom, the families of Billy Vachon and Tim Williams and blast survivors Tony Viall and Tom Sled. Not one of them expressed animosity toward this man who was responsible for the killing. It was the circumstances, not the person behind them, that had left them angry and grieving. So they felt compassion for Hesston rather than any vengeful wish that he answer now for what he'd done so long ago.

Sam Pilkinton, the Mississippi catfish farmer who had told me about being in the classroom, warned me not to contact Hesston. He said the man no doubt had spent thirty years trying to get over it and had suffered enough. If he were Hesston and got a call from me, Pilkinton said, he'd tell me to leave him alone or I'd face a harassment charge. Maybe he'll want to talk, I offered, but Pilkinton thought that was unlikely. In the end, I assured him I'd be careful. Still, his comments weighed heavily on me. What was the best way to approach Hesston?

As I considered what to do next, Dick Bielen helped me find out as much as I could about Hesston through public records, tracking his life after Vietnam through names, addresses and phone numbers. I thought about writing Hesston a letter and showing it to nurse Lynn Bedics, who had tended to Nicky. She agreed to read it with an eye toward how Hesston might react. I told her I'd welcome suggestions but that I was determined to contact him. Weeks went by. I didn't put pen to paper until Mary and I vacationed in the Finger Lakes region of New York the last week of August 2000. On the porch of a bed-and-breakfast, I wrote in part:

> It's been more than thirty years since that terrible July day, and no doubt you've spent much of that time—if not all of it—trying to put the horror behind you. Your suffering is more than I can begin to comprehend...

Days later, I dropped it off at the Allentown Veterans Affairs Outpatient Clinic for Lynn and soon heard back from her. She was glad to see my approach was sympathetic to Hesston but felt it was important that I say up front that I was writing a book. I changed one line to read: "I just want to know what happened, for the book I'm writing about Nicky."

Lynn spoke with a clinical psychologist at the VA clinic who counseled veterans with post–traumatic stress disorder. Psychologist Steve Teders warned that my letter might prove harmful to Hesston because there was no predicting how he'd react. He had, after all, caused the deaths of his own men. He might have deep-seated guilt as a result of the trauma and might even be suicidal. If I were to trigger his memory, the shock could distress him. If I didn't know his mental state, Teders said, I'd be playing Russian roulette with his brain.

I couldn't determine whether Hesston was getting help through Veterans Affairs. That information was confidential. I spoke with a psychiatrist at the Sacramento VA Medical Center, Dr. Tin Zar Shain. When I explained the situation, he spoke about similar cases and strongly urged me to leave Hesston alone. If I contacted him, that might trigger flashbacks, Shain said, and post–traumatic stress patients are fragile. They suffer from deep depression and anxiety.

Such comments from professionals who cared for veterans slowed me down. But I had to communicate with Hesston somehow, or at least try. So the question became not whether to approach him, but how. I never sent my letter and I did not call him.

Was the best way to go to California and knock on his door? See if someone, a middleman, would contact him on my behalf? I didn't know. I was stuck about how to proceed.

Dick Bielen had an idea. He knew of someone who might be able to help: Don Ray, an investigative reporter who had given talks on how to conduct difficult interviews. He lived in the Los Angeles area and worked at the *Daily Press* newspaper in Victorville, California. He was also a Vietnam veteran.

A call I made to Ray in March 2001 led to a series of long, coast-to-coast conversations for months—and my decision on how to contact Hesston. "This isn't going to be easy," Ray said, "[but] I wouldn't worry too much about his cracking. He lives with this every day."

Ray's first suggestion was to avoid deception. "The reason is that when you finally make contact with him, you could lose him in a second if he suspects you weren't completely honest."

Next, Ray said, forget the idea of a letter or a phone call. Neither would do. The only way would be a well-planned visit in person that would require me to put myself in his shoes completely. "You must 100 percent be in his mind and know, without doubt, what his position is. This is the bulk of the work. You absolutely cannot allow any friends, spouses, kids, relatives, neighbors or anyone else to contact him on your behalf. They will *always* misrepresent you, even if they don't mean to."

Ray said I wouldn't be able to get Hesston to talk by telling him it would somehow help the world, or that it would be cathartic for him. "That's something he'll be surprised to experience *after* you've done your interview with him."

So I made my decision by eliminating all other possibilities. I would try to come face to face with Hesston—no intermediaries, no communication beforehand. I planned a trip to California in the fall of 2001 to knock on his door. If it were during the summer, he might be away on vacation. Instead, I made it for the week before Thanksgiving, guessing he'd be home then. Mary and I would go together. We'd visit San Francisco and drive along the coast. That way, even if my attempt to see Hesston failed, we'd have something to enjoy—though I couldn't imagine having fun under those circumstances.

In April, Ray and I talked about how I should approach Hesston. I'd have to be non-threatening and non-judgmental. I'd have to present myself as both a writer and a relative of one of the victims who just wanted to know what happened. Mary should be at my side, a calming influence, and I should carry paperwork to let him know I'd come having done my homework. Most important, Ray said, is that what I say in the first thirty seconds after Hesston comes to the door should answer all of the questions in his mind. I should come up with a half-minute script, memorize every word, practice saying it in front of a mirror and imagine how Hesston would react.

I spent weeks writing my introduction and sought input from Ray, Mary, my colleagues at the newspaper, a grief counselor and the pastor of our church. Everyone had ideas. The grief counselor, registered nurse Lorraine Gyauch, said I would have to earn Hesston's trust by being truthful, letting him do the talking and listening to him attentively. I would need to convey that I was his companion.

Invite him, she said, to walk beside me and be my friend.

Chapter 14

A Dog's Bark, a Man's Voice, November 13, 2001

A week before Thanksgiving 2001, two months after the 9/11 terrorist attacks, Mary and I boarded a plane outside Allentown, the start of an eight-hour trip to Sacramento with a stop in Chicago. Gripped by queasiness, I realized not all of it was anxiety about what might lie ahead in California. National tragedy hung heavily over us, and our fears had been heightened the day before when a full passenger jet with its tail broken off crashed in Queens, New York, killing 265 people.

On our plane, I reviewed what I'd brought along—years of notes, Hesston's military records and the casualty files for Nicky, Billy Vachon and Tim Williams. I planned to carry them with me when we walked up to Hesston's door, as Don Ray had recommended.

Among my notes was the four-page, typewritten script for what I'd say when we saw Hesston. On hour-long walks during the past two weeks, I'd recited it over and over, and if I forgot what to say next, I would pull the script out of my pocket, check it and move on. By the time we left Allentown, I had it memorized. That did not assure that I would remember it under pressure when I came face to face with Hesston.

I agonized over whether I was doing the right thing, about how Hesston might react when we arrived at his house unannounced. My thoughts raced from one potentially nightmarish outcome to another, given that Hesston certainly would not want to again face the facts of the awful act he had committed. It was so potent in its senseless violence that he had probably relived it countless times in shame and horror. No doubt it still lurked behind every new day, intruding on his peace of mind, tearing away at his sense of self-worth. Such trauma could manifest itself in any number of ways when I tapped into it at his door. He might plunge into depression, become uncontrollably angry or hysterical, try to hurt or kill himself—or us. He might just slam the door in our faces, and that would be the end of it. And if he did talk, would he tell the truth? I'd heard and read that some veterans who'd suffered trauma create their own version of events to protect their minds from the painful reality of what they'd experienced. Hesston might tell me what he believed to be the truth but wasn't.

We landed at Sacramento International Airport and impulsively changed our plans. Originally we intended to drive first to Sacramento, where we had reservations at a bed-and-breakfast. We figured we'd be tired from a long day of travel. But when we landed, I found that I wanted to press on. I was too edgy to take it easy, knowing my reason for traveling 3,000 miles was less than an hour away. Beyond my impatience, I felt drawn to Hesston's hometown in a way I couldn't resist. Mary understood.

The day, Tuesday, November 13, was overcast and warm, in the low sixties. As far as we could see along the highway, the land was flat, far different from the rolling hills of eastern Pennsylvania. The sights were foreign to us—palm trees, lazy and peaceful with their drooping fronds, a soothing distraction. Mary remembered that she had been to central California as a teenager to visit relatives—her mother had grown up in the foothills of the Sierra Nevada near Yosemite National Park. She wondered aloud how hard it must have been for her mom to leave such a pleasant place.

The closer we got to Hesston's town, the tighter I gripped the wheel of our rental car. We followed the signs, turned off the highway and headed for the town hall. I went inside and bought a map of the town from the receptionist, a young woman. When I mentioned the street I was looking for, she said she lived on it and gave me directions. I wondered whether she knew the Hesstons, and asked what number her house was. She told me somewhat guardedly, but it wasn't close to their home. I said nothing more. She regarded me with some suspicion and would again within the hour.

It was a ten-minute drive to the Hesstons' street, a neat, tree-lined neighborhood of ranch and two-story houses. We cruised and counted the numbers until we found the right place. A car was in the driveway.

I parked on the street, a little up from the house. It was about three in the afternoon. Mary and I said a prayer for Hesston and ourselves and got out. I tried to summon courage and stop shaking. I fished around in my satchel for my papers, pulled them out, fussed with them until I got them in no particular order, as long as Hesston's Army file wasn't on top. I wanted to catch his interest, not make him angry. I knew I was stalling. I'd been imagining this moment for more than a year, and it was difficult to grasp that it had arrived.

Then we saw something that puzzled us. A teenage girl coming down the sidewalk turned into the driveway and went through a gate alongside the house. Mary and I looked at each other. Did Hesston and his wife have children?

We walked to the front door. I was trying not to think. I pressed the doorbell. The girl answered. The Hesstons don't live here anymore, she said. They moved last summer.

"Oh, no!" I blurted, and felt an avalanche of panic and dismay. In all of the scenarios I had tried to anticipate, this was not among them. I had checked Hesston's address before we left, and it hadn't changed. Had he moved too far for us to reach him on this trip, to someplace like Idaho or Colorado or even farther? It had taken me so long to find him. Would I now have to find him again?

Agitation swirled, fogging my mind to the point I felt almost immobilized. There was an instant, eerie sensation of being detached from time and place, cast away in a void. All my work, all my preparation had brought me to this place of nothingness. Yet as I teetered on the edge of this precipice, I happened to grasp something about what the girl had said. "They don't live here anymore" suggested she knew the Hesstons, and if she did, she might know where they had gone. I heard myself asking her calmly but insistently, trying to mask my anxiety.

Oh, they're still around, she said, nearby but on the edge of town. *Still here!*

The speed and depth of my relief were unlike any unwinding I could remember. From the highest tension, I plummeted to a warm and sublime state, as if I'd just gotten a shot of morphine—and a second chance. I exhaled deeply. Through the luck of meeting this friendly teenager as she came home from school to the house that used to be Wayne Hesston's place, the mission had been saved. The girl didn't know his exact address, but at least now Mary and I could go on.

We drove back to the town hall and approached the receptionist. I explained that the person we came to visit had moved to another street and could someone tell us the new address. She looked at me darkly and said no, she was not permitted to do that.

"Why don't you just call him?"

Of course we couldn't do that, but I didn't want to press my luck with the receptionist. She was suspicious, and the police department was directly across the lobby.

Mary and I went outside, and forgetting we had a cell phone, used a pay phone to call directory assistance, which is what we should have done in the first place. I was so nervous, it was a wonder I could push the right buttons. An automated voice gave Hesston's phone number, which hadn't changed. Mary suggested I call the operator and ask for his address. The woman promptly gave it to me. When we looked at the map, we saw that the girl at Hesston's former home was right. He had moved less than half a mile.

We identified the place from the mailbox. It was a wood-frame house with a car and a pickup truck on the property.

Mary and I parked across the street and again said a prayer. We asked God to be with Hesston and to help us say the right things. We walked onto the property with the same uncertainty that hung over us as we had approached the other house. We didn't know whether Hesston was home or what would happen if he were.

It was about a quarter to four. As we came up a hedge-lined path to the front door, I saw a light inside the house and heard a dog bark and a man's voice. He's there, I thought, and the jolt of excitement almost stopped me. Somehow, I kept walking.

On the porch, as on porches across the country, a large American flag hung, surely to commemorate the victims of September 11 and to support the emerging war on terrorism. Mary stood to my left, close to me. I opened the screen door and knocked three times.

In a few seconds, a man came to the door and pushed open the screen. He appeared to be in his early fifties, stocky and athletic. He wore a baseball cap, a dark hooded sweatshirt, blue jeans and white socks, no shoes. Under the brim of his cap was a hard-etched, clean-shaven face of natural toughness. He had blue eyes, brown hair neatly trimmed over his ears.

"Mr. Hesston?"

"Yes."

I was sure he noticed the sheaf of official-looking papers I was carrying, though his eyes never left mine. Barely containing the tension in my voice, I started saying the lines I'd committed to memory.

"My name is David Venditta. I'm a writer."

Chapter 15

"When you screw up in a war,"
November 13, 2001

Hesston and I shook hands. My knuckles creaked in his iron-hard grip. He withdrew his large hand and allowed the screen door separating us to close.

"This is my wife, Mary," I said, gesturing to her close beside me. "We came all the way from Pennsylvania today to meet you."

I'd caught his attention but he said nothing, only stared at me. My mouth went dry as I moved toward articulating why we stood before him.

"I'm writing about my cousin Nicky, a guy I really admired. He was an Army helicopter pilot. I lost him in Vietnam. I've wanted to talk to you ever since I found out what happened to him at Chu Lai."

There, it was out. In the space of a few breaths, I had taken him back to Vietnam to what was surely the most traumatic day in his life, the scene of his death-dealing failure three decades ago. He didn't stir behind the mesh of the screen. He stood statue-like, absolutely silent, his eyes fixed intently on mine. I kept going, searching his face.

"I'm standing here really afraid I'll upset you or cause you discomfort. That's the last thing in the world I want to do."

As I said it, I was on edge about what might be going on inside his head, heightened by a stab of panic that I might forget a thread of my message, irretrievably squander a precious opportunity born of years of labor.

The anxiety passed in seconds, fought off with the certainty that I was prepared for this encounter and would not, could not, mess up.

"I had thought about writing to you or calling you, but that seemed too impersonal. I'm here so you can see me. I want to be up front. You need to know I can't write about my cousin without writing about you. You're like him. He was a victim of the war, and it's clear to me that you were a victim, too.

"My heart goes out to you."

I felt good about saying that, about reaching out to him, and it was clear that Hesston was with me so far. He had not walked away from the door or interrupted me or broken our intense eye contact. My script was playing out the way I had rehearsed it a hundred times, and I was near the end.

"I don't know whether this means anything to you, but I want to say, on behalf of my family, that we've accepted it and moved on. We aren't angry with you. We don't wish you any ill will. We understand that what happened in that classroom at LZ Bayonet was just one of the many terrible things that can happen in a war. We wonder what went wrong with the system that something so awful could happen."

Hesston did not flinch at the mention of LZ Bayonet, which would convince him that I knew details. He hadn't moved or spoken since we'd shaken hands, and his expression remained utterly blank. But with just a couple of sentences still to say, I realized that the moment in which he revealed what he would do was fast approaching. My gut tightened as I finished.

"I'd like to be able to write that you are OK, too, like the rest of us. It's important for me to know that. Would you be willing to talk to me, a relative of one of the guys who was in that classroom July 10, 1969?"

An agonizing moment passed.

He pushed the screen door open.

"Yeah, come on in."

Mary and I turned to each other, astonished. We followed him into a dim living room, which had only the light of the late afternoon sun filtering through shaded windows. A worn couch on the linoleum floor faced the fireplace. Tabletops had coasters with scenes of hunting and fishing. He had a fire going, despite the mild temperature outside, and the room was uncomfortably warm.

He stopped and turned toward us. There was an awkward moment when none of us moved or spoke. I glanced at the couch. He gestured to it hastily as if he had forgotten about it, and told us to please sit down. Mary and I settled in chairs near the door—a precaution that the grief counselor had urged.

Hesston had been fixing himself dinner. Beef roasted in the kitchen. He excused himself to tend to his cooking. When he came back, he said we were lucky to catch him at home. He had just returned from a day and a half on the road in his truck, making a delivery to Los Angeles. He would be leaving early the next morning, Wednesday, on another trip and wouldn't be back for three days. So if Mary and I hadn't come when we did, if we had gone instead to the bed-and-breakfast in Sacramento as originally planned, we would have missed him.

Hesston sat down on a chair in the middle of the room, facing us. He spoke haltingly at first. "Something went drastically wrong, and I have to live with it all of my life. It was an ugly situation. I made a mistake, and when you screw up in a war, people get killed."

At first I scribbled his words in a pocket notebook but soon felt uneasy about it. He saw me, and it didn't appear to bother him. Still, I didn't want my note-taking to distract him, and I thought that when I was looking down to write, I might miss something in his demeanor, which was as dark as the room. I had ruled out a recorder as too aggressive and intrusive, wanting to be the relative hoping for answers, not the writer looking for a story. I put my pen and notebook down.

Hesston said he appreciated hearing that my family didn't blame him for what happened, but he rejected the idea that he was a victim like the "two or three guys" who were killed. (I made a mental note: He wasn't sure how many died.) He described himself instead as a victim of a cruel irony.

He had saved lives as a soldier in the field, he said, moving wounded men to medevac helicopters so they could be airlifted to a hospital. But that didn't count for anything after July 10, 1969, because he had thrown away that sense of satisfaction, of having done some good, when he let go of the grenade. He had rescued some men, he said, but that was canceled out when he took the lives of others.

Pain and regret were all too evident in what he was saying.

Hesston said he could not go easy on himself. He admitted he was responsible. But he pointed to other ways GIs died. What about the guys, he said, who mistakenly directed artillery fire onto their own men? They were responsible for deaths, too. Did they feel the weight of guilt as acutely as he did? Did they feel any guilt at all?

He waited for an answer that I could not give. I knew he was expressing hurt and anxiety with which he had been living for decades, but I was focused on my objective—learning details of the day Nicky got hit.

Yet I dared not let that impatience show. As eager as I was to find out, I knew from my experience interviewing war veterans for the newspaper that I shouldn't rush. Instead of barging ahead, I had to step back. I had to restrain the impulse to press immediately for the critical information, because that might work against me. It might intensify Hesston's anguish, causing him to withdraw, when what I really needed to do was ease his mind a little. If I could get him to feel comfortable talking, the hard parts would come easier. I'd start with something that wouldn't put him off and let him tell the story at his own pace.

I asked him how he had become an instructor.

He said he'd had the job with the Americal Combat Center for less than a month and had conducted twenty to thirty classes. Before that, he was an M60 machine gunner with the 46th Infantry. He had been in Vietnam since the fall of 1968 and had advanced in rank from private to sergeant. Early in '69, he said, shrapnel gouged his left arm and hand.

Those wounds, for which he was awarded the Purple Heart, were less troublesome than an injury to his left knee from a fall, he said. The knee swelled up and bothered him so much that he couldn't keep up with the other guys in his company. He hobbled to the rear, to the division's base at Chu Lai, on temporary stand-down. After his leg improved, he would rejoin his unit.

He said someone spoke with him about teaching new arrivals, a job that sounded appealing. He believed that with his experience in the bush and against the enemy, he could offer replacements sound advice on how to fight and stay alive. Besides, he thought some instructors "didn't know shit from Shinola." After sitting in on a few classes, he volunteered to give talks on grenades and claymore mines.

What was the point of the class? I asked. What was your routine?

The idea, he said, was that if you didn't respect these weapons, they'd kill you. Another sergeant had told him to fumble the inert demonstration grenade for the new guys. It was standard practice in the class, a built-in part of the instruction. "You've got to do this," the guy said. "You tell them they've got about five seconds, and that's when you fumble it."

Hesston said his morning class lasted forty-five minutes. He kept his box of props in the cadre room, a nook where inert claymore mines and other demonstration equipment were kept. His box held a couple of writing tablets and his demo grenade, which was in a fiberboard container.

"What kind of grenade was it?" I asked.

He was looking down and seemed to draw a blank. He said nothing.

"Was it an M26?" I offered, referring to the commonly used fragmentation grenade. Shaped like a lemon, it had a smooth, cast-iron body with a ridge around the middle and weighed about a pound and a half. Once the detonation sequence had begun, the smokeless time-delay element took about five seconds to burn down to a detonator, which acted like a blasting cap and set off a core of high explosive, bursting the grenade and spraying quarter-inch metal strips and bits of shattered iron casing. It had a blast radius of sixteen yards.

Still looking down, Hesston only shrugged.

He knew the grenade was his, he said, because he had notched the container. To make the grenade inert, he said, he had removed the fuse and detonator.

Thirty to forty minutes before each class, he went to the cadre room and checked the grenade by unscrewing the top and ensuring the absence of the fuse. When it was time for class, he carried his box of props into the classroom, stepped to the stage and faced forty to fifty guys who were new in-country.

Hesston said he never knew any of them. They were fresh-faced replacements, and he'd tell them things that might help to save their lives. His main message was: Always be aware of what's going on around you and never take the easy route, because mines, booby traps or an ambush probably await you there.

I asked him what happened on the day of the explosion.

It was just before his twenty-first birthday, he said. That morning, as he had done every day that he taught, he went into the cadre room, took the grenade out of the box and out of its container, unscrewed the top and confirmed the assembly was inert. He put the grenade back in the box, which he left in the cadre room. About a half-hour later, when it was time for the class to start, he returned to the room for his box of props. Then he entered the classroom and began his talk. After several minutes, he held up the grenade and talked about how dangerous it was, that you shouldn't fiddle with it because you only have five seconds before it goes off.

Hesston said he pulled the safety pin and deliberately fumbled the grenade onto the floor. The safety lever flew off. As the grenade rolled toward the benches, he said "oops," as if it had gotten away from him by mistake. Just as in previous classes, the guys reacted with little surprise.

They just looked at him. He didn't remember that anyone ran off or moved to shield himself, and that wasn't unusual, he said. When he was a new arrival taking the class and saw the instructor toss a grenade, he thought it was a farce. These guys in front of him probably had heard about the routine.

Here, Hesston hesitated. We had been sitting in his stuffy living room for more than an hour, fighting the dulling effect of the warm air, and now he was ready to say what I hoped to hear. He didn't look up, as if he had to avoid eye contact to pull the experience from memory. Pain creased his face. He drifted away from us, withdrawing into himself to conjure a flash in time he'd surely endured over and over again and didn't want to revisit, but felt duty-bound to convey.

Then he spoke.

The grenade rolled under a table up front. An ear-splitting explosion followed. Blood squirted out of his chest and neck. Smoke was all around. He collapsed to the tile floor and knew he was badly cut up.

"After that, I was in a daze."

And that was all he would say—a few seconds of fractured remembrance, delivered simply and directly as plain fact, without visible emotion. It seemed like such a spare description of an event I'd been chasing for years—the moment of an indoor explosion of such force that it instantly killed Tim Williams, tore open Nicky's legs and those of Billy Vachon, Tony Viall and Tom Sled, cut into several dozen others and filled the classroom with smoke and the shrieks of wounded, stunned and panic-stricken men.

In all the times I'd played over in my head how hot metal fragments traveling at blinding speed had felled Nicky and the others—a picture I'd gotten from my talks with survivors Tony and Tom—I had no image of the explosion's impact on the instructor. I didn't know whether he had fallen to the floor or stayed on his feet, whether he was seriously hurt or merely scratched. Tony and Tom couldn't say because they hadn't seen him, or didn't recognize him, immediately after the grenade went off or as medics tended to the wounded outside the building.

Now, finally, I had something of a picture. When Hesston mentioned the squirting blood, he'd bowed his head and motioned with his hands under his chin to show how the fluid shot out in a spray of red. But he didn't give any other details that would present a more complete image—maybe he didn't have them to offer—and I didn't break in on him. I had allowed him to speak at his own pace, and it appeared that he had spilled the words in a short burst to get past the remembering.

"Well, that's about all I can tell you about that," he said, his eyes cast downward. "I hope that answers your questions."

Mary and I had said little. We hadn't needed to prompt him with a lot of questions because he talked freely, as if unburdening himself of long-held guilt. I felt more at ease and hoped to connect with him in a more personal way, so I asked him about himself.

"Where did you grow up?"

"Chico," he said.

"Chico," I repeated. "Like one of the Marx Brothers."

I had meant it to lighten the mood, but instead it threatened the fresh and fragile link we had with him. Hesston's face darkened, he lowered his chin and glared at me. "I don't see what's so funny about this!" he burst out. "We're talking about something very serious here!"

Paralyzed, I quickly turned to Mary, who looked at me wide-eyed and afraid. Suddenly, the warm room turned oppressively hot. That one stupid comment in this delicate setting, where I should be letting Hesston do the talking, could end our meeting and ruin any chance of learning more from him.

I fumbled with words about not meaning to make a joke and braced for whatever might happen next.

Nothing.

Hesston's flash of annoyance passed. An urgent warning to myself: Shut up! *Just shut up!*

Recovering, I asked what happened to him after the grenade went off.

He said he was hospitalized at Chu Lai—though he couldn't remember at which of the two hospitals there—evacuated to a hospital in Japan, and then sent to the States. At Fort Lewis, Washington, he was interviewed several times about the explosion. He told his questioners, including a colonel or lieutenant colonel, about his routine and said he was at a loss to explain why the grenade had been live.

They had checked his records, he said, and found no red flags that would indicate he was anything but an upstanding soldier. They called him back a couple of times, but nothing more came of the investigation, as far as he knew, and the matter was dropped. He said no action was taken against him.

Apparently the Army had not followed up with a meaningful investigation of an extraordinary occurrence: three soldiers' deaths at the hands of a fellow American. No one—not even Hesston—had ever gotten a forthright explanation of what happened.

Chapter 16

Making a Case for Sabotage, November 13–17, 2001

Mary and I had been sitting with Hesston for longer than an hour. I can't say exactly because we were focused on the man we had crossed the country to see, this onetime soldier it had taken me years to identify. We had made it through his door, survived a misstep that sparked an angry outburst and remained in the room, gently following his lead. We'd hung on his words as he slowly, haltingly, spoke. We'd seen that our surprise visit had shaken him and how he was struggling to accommodate us. We'd seen the pain that consumed this fifty-three-year-old as he remembered a tragedy that happened when he was barely out of his teens.

I felt some relief. I thought it was brave and decent of Hesston to go on talking, that it was right to have this airing of his account and that it ultimately would be good for him as well as for me and the families of the dead.

Still, there was a prickly tension. We had only just met him and couldn't know how he might react in an hour or the next day or a week from now. The air in Hesston's already uncomfortably warm, fireplace-lit living room thickened and closed in on us, making us drowsy after our 3,000-mile journey. Would we miss something he said? A key detail? *Stay alert*, I told myself. And with late afternoon giving way to twilight, the room grew darker, reflecting Hesston's mood as he faced a relative of one of the men he had killed, and as he dwelt on his deed of July 10, 1969.

Again, he said he accepted responsibility for the explosion. It wouldn't have happened, he said, if he had checked his props to make sure they were safe for the classroom. But instead of stopping there, he let us in on a scenario he clings to—an explanation embraced by others I'd spoken with but unsupported by any known evidence and ruled out by the Americal Division, which had reported no incidents of sabotage in that quarter of 1969.

With all his heart, Hesston said, and to the day he dies, he would believe that the Viet Cong were responsible.

Someone had made sure he had an armed grenade—a person who knew his routine. He said anyone could have wandered into the cadre room of the orientation building in the half-hour to forty minutes before the class and put a fuse and detonator into his grenade, which was in a fiberboard container he had marked, or substituted a live grenade. It would have taken only ten seconds.

Who would have done that? I asked.

"There were plenty of Arvins around," he said, using the nickname for South Vietnamese troops. I later determined he was right about that. Two soldiers who were with the 1st Battalion, 52nd Infantry, based at LZ Bayonet, confirmed there were ARVN soldiers on the site in 1969. Troops in the 2nd ARVN Division operated in the region, as well as local military groups called Popular Forces and militia units, the Regional Forces. Other Vietnamese were ever-present, doing all manner of work such as cleaning hooches and doing laundry.

"There's no doubt in my mind that somebody tampered with that grenade," Hesston said.

He said he hoped it was a Vietnamese. He couldn't fathom why a GI would want to kill him or other Americans. He'd been at Chu Lai only a few weeks, not long enough to know anyone well, much less to draw that degree of hostility. Besides, someone who wanted to kill him could have fragged him with a grenade while he slept, or shot him. His big mistake, he said, was in assuming that the orientation building stood in a secure area. To a soldier who had spent months in the boonies, LZ Bayonet was a haven from constant danger. He thought it was relatively safe.

"I didn't even carry a weapon, not even a pistol," he said.

As I heard this case for sabotage, I realized Hesston knew little about what had happened that morning, not even how many soldiers died from the explosion. He had been in the dark all these years, torn up inside and left to wonder. The Army, in apparently not following through on an investigation and reporting on its findings, and in filing no permanent record, had left the root cause of the blast open to speculation. As a result, in the absence of certainty about what happened in the classroom at LZ Bayonet, Hesston had come up with his own conclusion: that the horror had been a deliberate, cold-blooded act.

He paused, turned away from us, looked down and said, as he had before: "I guess that's all I can tell you about that."

I saw an opportunity to just have some small talk so we could get to know him and he could relax a little. This time I was determined not to say anything flip that might upset him. Instead I asked him to tell more about himself—about where he grew up, where he went to school, what he did.

But the questions flustered him, and he erupted.

"Why do you want to know that? What does that have to do with this? That doesn't have anything to do with this!"

The heat from the fireplace pressed on me; it felt suffocating. His outburst left me numb and worrying about what might happen next. I wondered, too, how he could sit there in his sweatshirt, still with his cap on, in this far-too-toasty room. I groped again for the right words, telling him honestly that I just wanted to know more about him—who he was and where he had been. Just as before, the shaky moment passed. Hesston's dark visage disappeared.

In asking him about himself, even though it was during a lull in our conversation, I had apparently interrupted him. I had not picked up on his strange rhythm, a stop-and-go jerkiness to his storytelling. He seemed to be locked onto, and grappling with, his memories of Vietnam and the explosion. And though he seemed in the present, he was deep in his own world and Mary and I were intruders.

Once again he started talking of the long-ago time in Vietnam, bitterly volunteering his thoughts about the war. It had been waged for no good reason, he said. It was a waste, ill-advised and ill-fought under terrible leadership. He said GIs who should have had two years of training instead of six months were treated like toys and moved around like pawns in a chess game. But they did their jobs well.

He spoke affectionately of the guys in his company and their leaders. He felt that he, "a simple farm boy," was blessed to be with them. They were good soldiers who knew how to fight and stay alive under the most frightful conditions, and who stuck together and would never leave anyone behind. They developed a bond of trust; they were smart and savvy. They didn't act rashly and rush into ambushes or stupidly expose themselves to the enemy. As a result, there were few deaths among them compared with other, poorly led units.

One of his fondest memories, he said, was of returning to his company from Da Nang, where he had been hospitalized for a while, delirious with fever. He didn't like it there among people he didn't know.

"I always wanted to be with the guys," he said.

Gradually, Hesston said, he developed a keen interest in the war as time separated him from the trauma of the explosion. His perspective changed. He had come to admire the former enemy and the people at home who wanted us out. He said he now respected the Vietnamese communists for defending their homeland tenaciously and with whatever means they could muster. They were resourceful and intelligent. "We were amateurs fighting pros," he said. Most startling, he credited the antiwar protesters with cutting the war short and saving at least 20,000 American lives. At first, he said, he hated the protesters. In 1968, when he was at Travis Air Force Base in California bound for Vietnam, they were there, spitting at the GIs. But then he came around to believing that these troublemakers were right about the war—and stronger-minded than he had been. His father and stepfather had been Marines. When he was drafted, he said, he believed that serving in the military and going to Vietnam were expected of him. He felt it was his duty.

But he came to understand that hundreds of thousands of Americans answered another call, which told them the war was cruel and unjust. Some were so adamant about not serving that they were willing to sacrifice their freedom and go to jail. In all, 8,750 were convicted of draft evasion. About 10,000 left their homes and families for refuge in Canada. Their refusal to do the government's bidding, along with the massive antiwar protests across the country, eventually proved too much for Washington to ignore. Hesston said he thought that was a good thing because ultimately it meant no more young men would die in Vietnam.

He didn't tell us in any detail about his own homecoming in the summer of 1969. He would have been recovering from fragment wounds and distraught over the horror he had unleashed in the classroom. He would not speak of the specialized treatment at the hospital at Fort Ord that his records show he received. Instead, he stuck to generalities, saying he knew a lot of guys who were "messed up" long after their return. When they sought help at Veterans Administration clinics, they became dependent on drugs to feel better. That was not for him. When I asked what the VA did to help *him*, whether he received counseling, he wouldn't talk about it. He said only it helped to talk things out.

Still, he said, he confided in few people about what happened, and he needed a crutch, which was alcohol. He drank to soothe the hurt. He played semi-pro baseball, drove tractor-trailers on long hauls and got married. He said his wife and a longtime friend told him that he couldn't go on torturing himself. He had to lay off the drinking, pick himself up and get on with his life, because there was nothing he could do to change the past.

Over time, he came to accept that. An illness his wife suffered helped him see the need to bounce back for her sake, and he found an outlet in coaching a high school baseball team.

But he said the bloody classroom was never far from his mind. The deaths and injuries to the soldiers happened, in part, because Hesston believed he was in a secure area and his props were safe. That trust was misguided, and he was determined not to allow himself to be blindsided again. He said he'd taught himself that where safety was concerned, he could never presume that someone else has ensured it. He must be the one. He would occasionally haul loads of explosives and hazardous materials. Before a trip, at night as he lay in bed, he mentally ticked off the safety checks he needed to make on his truck before he took it on the road. Even if he was told something was already done, either mechanically or in securing a load, he insisted on examining it himself. It's what he had to do for peace of mind. He didn't want anyone else to die or be injured on his account.

The torment over the grenade was not the only scar he carried home from the war. Once, he agreed to speak about Vietnam in a school. A twelve-year-old girl asked him what it was like to kill women and children.

How did he respond to that? I asked him.

"The way I am now," he said, his eyes glistening with tears.

As if the emotional scars weren't enough, Hesston still suffered from physical ones as well. Malaria that sickened him in Vietnam returned to give him episodes of fever and blisters on his skin.

I felt for him when I heard the price he has paid. How sad that after thirty years the war still, from time to time, bedeviled his mind and degraded his body. I understood that his sacrifice was like that of so many others who have served in uniform, who, long after leaving the battlefield, lead difficult lives. For some, the grave beckons much too soon.

In the growing dusk, Mary reminded me that I had brought an email from Tim Williams' sister, Jill Bell, a teacher who had a message for Hesston. I told him Williams was one of the *three* young men who died from the grenade. I leafed through my papers, pulled out Jill's email, put on my glasses and held up the printout to catch as much of the twilight as I could from the window behind me. Even though it was growing dark, Hesston had made no move to turn on a light.

"I sincerely hope that the instructor is doing well and has put this terrible accident behind him," Jill wrote. "Thirty years is a long time for anyone to carry the weight of regret. It happened. It was unintentional. It's time to bury the pain.... Please tell this soldier that I send along a hug of compassion and forgiveness."

Hesston just looked at me absently.

Yet again, he said he couldn't think of anything more to tell us. This time, we all stood up, but he started talking again, and we stood facing one another for perhaps a quarter hour. We went out onto his porch in the growing darkness. He glanced at the papers I carried at my side and said he guessed I'd seen Army records about what happened. But he didn't ask to see them and said nothing more about it. He never asked how I identified and found him.

I told him that we wanted to stay in touch and gave him our cell phone number, and then we left to spend the next several days in Sacramento and San Francisco. On Friday afternoon we called Hesston, who said he'd been trying to reach us and wanted to see us again the next day at our motel near the Sacramento airport. It turned out he had left several messages on our phone, one in which he groused because I hadn't called back. "I don't know what your problem is," he barked. "Maybe you don't know how to use the phone." Ouch. It was our first cell phone, clunky by today's standards, and we had gotten it specifically for this trip. We had made a mistake in not taking it with us while sightseeing.

In our motel room Saturday afternoon, Mary got caught up in the movie *Back to the Future* on TV, but I wanted to get down to the lobby and wait for Hesston. I'll go, I said, and see you later. "No," Mary said, getting up, "I want to be there if he kills you."

She was only half joking. We didn't know Hesston's state of mind. He'd had a few days to think about why we'd come to see him. Maybe it had built up in him to the point where he was no longer rational.

Hesston showed up in the lobby in blue jeans, a T-shirt and the cap we had yet to see anywhere but on his head. The three of us walked to a Burger King and sat at a booth, and Mary bought him a Coke.

"If I didn't trust you, I wouldn't be here," he said, "and I have nothing to hide."

I put my microcassette recorder on the table and turned it on, saying, "This thing is running." He shot a look of suspicion at it but didn't object. Around us was the din of a fast-food restaurant doing brisk business on a weekend afternoon—kids squealing, parents chirping, laughter and workers calling out orders. I asked Hesston more questions now, and we went over the same ground we had covered before.

To jog his memory, I gave him a copy of the article I wrote for the Americal Division newsletter in early 2000, a diagram I'd drawn showing what happened in the classroom, based on accounts from Tony Viall and Tom Sled, and a photo of the orientation building as it looked in 1969. He seemed fascinated by everything but handed the photo back.

I asked him questions I hadn't raised before about the grenade routine. He said it was a "known training deal," a common practice in the classroom before he'd become an instructor. But he couldn't tell me whether higher-ups knew about it or whether they condoned it. Nor could he recall who supervised the class or if anyone supervised it, or who his superior was.

But he repeated his defense of himself.

"I stand by my record as a good soldier," he said. "I know it was not my fault that grenade got in there. Somebody tampered with it. Somebody got to the system and knew what was going on."

Still, he accepted responsibility for the deaths of Nicky, Billy Vachon and Tim Williams and the injuries to dozens of others. He had to live with what he'd done, he said, and it racked him for a long time. In the end, some good emerged from his anguish.

"I had to pick myself up and say: I better lead my life as exemplary as I can, because nothing is going to change what I did."

As we said goodbye in the motel parking lot, I told him how I'd agonized over how to contact him—by letter, phone call or in person—and that it seemed the best way would be to come and see him without telling him in advance.

"You did the right thing," he said.

I told him we were amazed that when he answered his door and learned why we were there, he didn't seem shocked and he invited us in without hesitation.

"I *was* shocked," he said, "but I figured if you had the guts to show up on my doorstep, I had the guts to talk to you."

With that, he walked to his pickup truck and drove away.

Mary and I flew home elated. We had done something extraordinary. Dire predictions about how Hesston would react turned out to be unfounded, and he had seemed all right, even eager to talk to us. All the preparation I'd done had paid off. All the people who helped me had been on the mark. Most important, I now had Hesston's account of what happened as he remembered it, or allowed himself to remember it. At the very least, his version raised questions about the Army's decision to label the explosion an accident, a determination not explained in any military record of the incident I was ever able to find.

A week after our homecoming, I called Hesston to see how he was doing. His wife answered. She had not been at home when Mary and I came to see him, so we hadn't met her. I asked if her husband was there. Yes, she said, and I heard her say to him, "Don't you want to talk to him?" Then he was on the line, terse and grumpy.

In a few minutes he loosened up and I understood his annoyance. He had read my Americal Division article and said it had inaccuracies. He said the guys I quoted were wrong in saying he had an entire box of grenades in the classroom. "That didn't happen. We didn't do that," he said. He had no Vietnamese aide, and there was no floor switch for setting off a charge outside the orientation building to scare the class. "I guarantee you there was nothing like that," he said.

There was just the grenade.

During the following weeks, the results of my contact with him resonated with my family and the Vachon and Williams families. Louise Vachon, Billy's mother, was moved to write to Hesston in December 2001:

> I'm sure you have wondered how the families and victims have dealt with the tragedy over the years. Speaking only for myself, I can assure you that there is no resentment or ill will towards you or anyone else associated with the incident. Everyone deals with tragedy and loss on their own terms, as I'm sure you have experienced yourself. In closing, I would also like to extend my heartfelt thanks to you for answering many questions that have remained unanswered for so many years. The trip David made to see you in California not only served as a closing chapter in his book, but also in my own.

Hesston did not respond. Nor did he write to Aunt Bert, who had also sent him a letter of forgiveness, or to Jill Bell, Tim Williams' sister.

They didn't seem to mind.

Chapter 17

The Marine Corps Way,
March 28, 1967

Just under twenty-eight months before my cousin met his fate at Chu Lai, a group of twelve Marines fifty-six miles away near Da Nang gathered on a typical sunny, warm March morning with their second lieutenant and staff sergeant outside a saw shop for a lesson in mine warfare. Their sergeant had been taught to handle mines, but the lieutenant and the others had no training. Nevertheless, they'd been picked to learn the tough task of clearing a minefield, an important skill in territory bristling with hidden death. They were kids, like most who served in Vietnam, the youngest eighteen, the oldest twenty-two. Most had been leathernecks less than two years, one for only seven months. Just kids, four of them black, eight white, from places like Beaver Dam, Kentucky; Pottstown, Pennsylvania; Scottsbluff, Nebraska; and Baltimore, Maryland. All would die that morning. Their lieutenant would die with them. The sergeant would live, but barely.

The Marine investigation would be swift, merciless and meticulously documented. Forty-seven years later, when I requested the documents, they were produced within weeks. The pages were filled with detail, brutal and clear, even disclosing disagreement among commanders as the investigation worked its way up the ranks.

What was impossible to get from the Army after decades of effort came so quickly from the Marines, it left me numb. No hemming, no hawing, no mystery, no fog of war. Just facts, starkly stated and complete. And a recommendation that, if heeded, would have saved the lives of Nicky and two fellow soldiers.

In all the years I'd been working on Nicky's story, engaged in the fruitless search for a written record of what happened to him, I'd wondered if there had been a similar fatal training accident involving U.S. servicemen in Vietnam, and if so, how it had been handled.

I had long ago concluded I would not find what I was looking for. No books I knew of had any such accounts, and I'd found nothing on the Web. Vietnam casualty databases list the causes of non-battle deaths—for example, a vehicle accident or drowning—but they don't specify whether they happened during training. Finally, in 2014, almost twenty years after I had started searching, a fellow writer gave me the break I needed.

George Lepre of New Jersey, who had helped me in my records search and was the author of the 2011 book *Fragging*, knew of an incident involving detonation of a live mine during a training session that resulted in multiple deaths. And he knew the surname of one of the dead—Bekiempis, from Bayonne, New Jersey. As soon as I had that basic information, I sought out online casualty records, which revealed that Lance Corporal Thomas C. Bekiempis had been killed March 28, 1967, in a non-hostile incident in Quang Nam Province, in the northern part of South Vietnam.

Now I could search for a report on what happened to him and the others. Using Bekiempis' name and his birth and death dates, I wrote to the Navy's Office of the Judge Advocate General at the Washington Navy Yard and asked for a copy of the investigation into his death. To my amazement, given the put-offs and runarounds I was used to in dealing with the Army, I got a phone call two weeks later from the Freedom of Information Act coordinator in the JAG Investigation Branch. He already had copied the file for me, seventy-one pages, and in summarizing the contents said that thirteen of fourteen Marines in the mine class near Da Nang had died in the explosion and that the only survivor was the instructor, who had thought he had rendered the mine inert. The JAG coordinator said if I wasn't satisfied with the quality of the reproduction that came in the mail, I shouldn't hesitate to call or send him an email.

The report was everything I thought it would be, handled exactly the way I'd think an investigation would have been carried out after the deaths of Nicky, Billy Vachon and Tim Williams. It was broken down into findings of fact (the details of what happened), conclusions and recommendations. The file included the investigator's statement on how he did his work on the case, the names of all the dead and wounded and of all the witnesses, a transcript of an interview with the badly injured instructor, a sworn statement he gave later, notes on the progress of his recovery, his service record, statements from witnesses, photos of the accident scene, a diagram of the type of land mine used in the class, a diagram of the scene showing where each Marine was standing when the mine went off, the death certificate for each Marine. It showed the investigation report was passed up the chain of command for all to see and comment on, from the battalion level all the way to the commandant of the Marine Corps.

The investigator, Captain John W. Dougherty of Headquarters Company, 7th Engineer Battalion (Reinforced), sent his report to the battalion commander six days after the fatal accident, beating an April 6 deadline the commander had imposed.

The findings of fact are numbered and each contains just one piece of information. For example, one says only that the mine warfare class was being conducted "in preparation for a mine-clearing project." Another says that a live M16 antipersonnel mine was being used as a training aid. The next line says that a fuse was inserted in the mine. After that: The fuse was rigged with two trip wires. Then, a staff sergeant passing by the saw shop at the company command post, where the class was being held outdoors, asked the instructor if the fuse was "deactivated," and he said it was. Then, "some force or forces unknown activated the fuse either by pressure on the pressure prongs or by pull on one or both trip wires." Next line: The M16 mine detonated. A Vietnamese employee of Company A who was walking past the saw shop was wounded. "Nguyen Huong received shrapnel wounds in the throat, abdomen, left leg and right elbow." The dead and wounded at the site, about three miles from Da Nang Airfield, were evacuated by helicopter thirty-two minutes after the accident. The report goes on to describe the damage caused to the saw shop and a truck that was hit by fragments.

The instructor, Staff Sergeant Jackie W. Kinslow, had been schooled in mines and was experienced in handling them, the report says. The officer in charge, Second Lieutenant Glenn M. McCarty, who died with his men, had read the technical manuals pertaining to the M16 mine and had discussed the training and the mine-clearing project with a supervisor, a captain.

After those findings of fact is a section labeled "Opinions." In it, Dougherty says the captain had done an adequate job of supervising preparations for the class. But the sergeant, Kinslow, had made a critical mistake. He thought he had deactivated the fuse by taking it behind a hill and blowing off the detonator with a blasting cap, but the fuse was still capable of firing when he inserted it into the mine. Dougherty concluded that the whole idea of using the M16 in the class amounted to a crime, since even someone experienced in working with mines could have easily made such a deadly error. McCarty and Kinslow, he said, were "grossly negligent in using live ordnance for training purposes." This was not a case of an accident happening in the line of duty, Dougherty said; it was misconduct.

There it was. A finding. A conclusion. In writing, preserved in the permanent record. Nothing like it exists for Nicky and his comrades. No record even enumerates how many died and how many were injured. No eyewitness accounts were kept and nowhere does the name of the instructor who threw the grenade appear, nor are there diagrams of the death scene. Missing too are memos from the high command grappling with how to handle the aftermath of the tragedy.

All of that appears in the Marine records on the mine incident. Holding the sergeant accountable, Dougherty recommended that Kinslow be court-martialed for "involuntary manslaughter by culpable negligence." He said battalion personnel should be required to attend Division Mine Warfare School before working with mines "in any capacity." And he recommended that the battalion commander establish a mine warfare school to supplement the instruction.

Lieutenant Colonel Frank W. Harris III, commander of the 7th Engineer Battalion, endorsed Dougherty's report, including the court-martial recommendation. Harris had this to add: "The handling of mines and firing devices are no safer than the individual using them. The combination of events that were initiated by Staff Sgt. Kinslow in the preparation for instruction could lead to no other result than the disaster that occurred. The use of a live mine … violated every safety principle and good judgment that is required when handling live ordnance."

The report was passed up to the commander of the 1st Marine Division, who also endorsed it, then to the commander of the 3rd Marine Amphibious Force, which oversaw all U.S. activity in the I Corps tactical zone, the northern part of South Vietnam. But that general, Lewis W. Walt, did not agree that Kinslow should be court-martialed. Walt said the evidence showed Kinslow "honestly believed that he was dealing with a fuse rendered inert by his precautionary action" before the class. So what happened in the class wasn't misconduct because it wasn't intentional, and there was no "clear and convincing" evidence that Kinslow demonstrated a reckless disregard of the consequences.

After Walt had his say, Dougherty's report went on to Victor H. Krulak, commanding general of Fleet Marine Force, Pacific. "Court-martial action against Staff Sgt. Kinslow is not considered appropriate in view of the insufficient showing of culpable negligence," Krulak wrote. But he directed his subordinates at the 3rd Marine Amphibious Force and the 1st Marine Division "to ensure that continued emphasis is placed by all commanders on safety and precautionary measures to avoid needless injuries to personnel or damage to equipment."

The Marine Corps commandant signed off on Krulak's memo, and the judge advocate general agreed with the top Marines, ruling that the mine blast was not the result of misconduct on Kinslow's part. He was not court-martialed.

All of this is easily accessible to the public at the Navy JAG office and can be copied free of charge and mailed within a few weeks of a request.

Online I even found a separate Marine document that noted the tragedy, a command chronology for March 1967 prepared by 7th Engineer Battalion headquarters. This is a rundown of the battalion's activities, similar to the Americal Division's Operational Report/Lessons Learned. The 7th Engineers document notes that at 8:20 a.m. March 28, "M16 mine detonated during training resulting in 13 Marines killed and one wounded." The Americal Division's report for the period that included the grenade explosion makes no mention of it.

Another online search revealed that the mine blast made Page 1 in newspapers across the country. The only mention of what happened to Nicky, Billy and Tim was in local stories reporting their individual deaths. There was never a media account saying that a grenade training accident at Chu Lai had left three soldiers dead and others wounded.

The enormity of the mine accident, the sadness it must have evoked for so many families, seized me. I kept going over the JAG file, and on one of those readings got a shock of recognition. It was the name "H. Nickerson, Jr." at the bottom of a memo. At the time, Major General Herman Nickerson, Jr., was commander of the 1st Marine Division. In March 1969, he took command of the 3rd Marine Amphibious Force. That force had authority over the Americal Division at Chu Lai.

Nickerson had been Army commander Lloyd B. Ramsey's boss— Ramsey, the old warrior Mary and I had visited years earlier in piecing together the story of the grenade accident. During our 1998 visit, he told us that when he heard about the blast at LZ Bayonet, he called Nickerson to inform him of the tragedy. Ramsey said Nickerson told him it was a terrible thing to happen but to let his subordinates handle it.

The Marine general had seen the report of the 1967 investigation, endorsed its strong language calling for a court-martial of the instructor, and forwarded it to the next-highest general. This gave me a direct link between the thirteen Marine deaths outside Da Nang and the deaths of Nicky, Billy and Tim more than two years later at Chu Lai. But it was too late to ask Nickerson about it; he died in 2000.

I had these questions: Did the Marine Corps and the Army talk to each other? After thirteen Marines were blown up in a mine-training class and the Marines agreed it was far too dangerous to use live ordnance in training, did that lesson get passed on to the Americal Division? Would that catastrophic loss of life demand that safety procedures throughout the military be revised?

Not necessarily, some said, and gave disheartening reasons. A military historian told me that the Marines wouldn't have bothered passing on such practical safety information because it was simply common sense. Like the mine accident, the grenade accident that killed Nicky was obviously preventable, wrote John S. Reed, an associate professor of history at the University of Utah and an Iraq War veteran. It was an example, he and others said, of how "shit happens"—a catch-all for when things go wrong. A document I read online offered a reason communications between the Army and Marine Corps might have been less than ideal: Marine commanders in northern South Vietnam had struggled to manage the Army's deployment in their tactical zone—a move aimed at helping them fight the enemy. "The command-and-control structure became overburdened," Major General George S. Eckhardt, who led the Army's 9th Infantry Division in Vietnam, wrote in a 1973 study.

It seems to me that if the Marines' lesson about the use of explosives in training had reached Chu Lai and been heeded, Sergeant Wayne Hesston would not have brought an M26 fragmentation grenade into his safety class on July 10, 1969.

Nicky, Billy and Tim would have lived.

Chapter 18

Change of Heart,
July 20, 2013

At the start, I wanted nothing more than to identify the man who killed Nicky.

I believed this faceless, nameless person had made an egregious error in judgment, yet the Army had not charged him with any crime or disciplined him in any way. It had let him go unscathed, never holding him accountable for the loss of life he caused.

For years, I did not question my conviction to publicize details that cropped up in countless documents and interviews. Withholding information is anathema in my profession. You don't sweep the truth under a rug; you shake it out and hang it on a line in the sunshine. In my early days on one newspaper, I had a zealous editor who regularly reminded his young charges that our job was to comfort the afflicted and afflict the comfortable. We had to be hard-nosed, probing, unshakable.

Then, in 2000, I encountered a man who had served side by side in combat with the instructor. A slip of Doc's tongue gave me the name I had pursued for years. But the conversation also planted seeds of empathy. Doc put a human face on the man I wanted punished, and he established the instructor as someone with soldierly knowledge, wits and courage under fire.

A year and a half later, when Mary and I finally met the instructor at his home in California, we were able to see him as a person haunted by the enormity of the error he committed when he was not quite twenty-one years old. We saw and heard the trauma in his words and demeanor, the stabs of panic that made his temper flare.

Over the years, he and I stayed in touch sporadically, in brief communications. I wrote him a long letter in the summer of 2013, as I was well along in the writing of this book, to tell him what I had learned since we had met. I wanted to be up front with him. I said it might be hard for him to believe, but I had never found a written record of what happened, only snatches of information. I told him what was in the daily staff journals and the condolence letter sent to Nicky's dad, and what some Chu Lai soldiers had told me—including the Americal Division commander, Lloyd B. Ramsey. I told him it wasn't clear who had conducted the investigation into the accident.

His response on July 20, 2013, was jarring. He called and left an angry voicemail message that reminded me of the harrowing moments when Mary and I sat across from him and he had erupted at seemingly inconsequential comments. He accused me of looking to blame someone and said I didn't understand about the war, that "stuff happens" and that's all there is to it. He said if I had shown up at his door and presented the results of my interviews with commanders and others—as I did in my letter—"I'd have run you off."

"I hope to never hear from you again!" he blurted.

His voice conveyed a displaced sense of fear and guilt he could not face. He did not sound rational. It worried me. I felt I had done the right thing in laying out the result of my work so he could know what I'd learned, but it had backfired. How should I deal with this? I consulted Don Ray, the journalist who prepped me for my visit with the instructor a dozen years earlier. Don suggested I wait a while and then write to him again.

For a year and a half, I did nothing. Then early in 2015, I wrote along the lines that Don had proposed. The letter said in part:

> I am writing to you to thank you and, most of all, to tell you that you were right. When Mary and I were with you in 2001, talking about the tragedy that happened to you and my cousin Nicky, you said something so simple that I did not appreciate it at the time. "Stuff happens." I made note of that, but it didn't really sink in. Not having been in a war, not even being a veteran, I couldn't grasp the idea of how things can go wrong, and how often they do. In a phone message to me on July 20, 2013, you said it again: Stuff happens. In the past dozen years, I've interviewed many more war veterans for the newspaper and gotten first-hand accounts of how happenstance made a difference and often proved deadly.... I was beginning to get the picture. But what really got me, what really drove home the point, was an occurrence I learned about several months ago while doing more research for my book. It was a training accident involving Marines outside Da Nang in 1967.

I told him what happened in that class, how the instructor had thought he had deactivated the fuse for an M16 mine. He was wrong; it could still fire. And that the mine exploded, killing thirteen Marines, and only the instructor survived. "Now, after all this time," I wrote, "I under-stand what you were telling me."

He did not respond.

His silence was the final step toward a decision I never would have thought I'd make through all the years of my search for answers in Nicky's death. I decided not to use the instructor's real name. And I have left out some details to protect his privacy.

Putting forty years of journalistic training aside was difficult. Yet I no longer believe that the right and proper path is to expose a man still troubled by the deaths he caused while he was trying to teach fellow American soldiers how to stay alive. I can't do something that could add to his suffering.

His name mattered to me. But now, so does he.

Chapter 19

"A man who did not run," 2015

The flight helmet that sits on my desk at home looks much as it did forty-six years ago. Its smooth, hard plastic in the Army's standard color, olive drab, shows no damage. The clear visor remains unscratched, the communication gear and wiring intact. Headset and foam-pad lining are still in place, though the padding is frayed with age. A blue label affixed to the front, just above the visor, bears the name VENDITTI.

Nicky wore the helmet at the controls of a Huey helicopter, and it is one of many reminders of his life that I keep and treasure. It came home from Vietnam to his parents in Malvern six weeks after his death, among about fifty of his belongings shipped from the Army's Personal Property Depot in Saigon. The helmet and other items are listed in Nicky's casualty file and range from the mundane—a pocket-size Army sewing kit of needles, buttons, thread, and safety pins—to the intimate—photos of his family and fiancée that he kept in his goatskin wallet. Separately, Uncle Louie received a U.S. Treasury check for $254.80, the amount Nicky had with him.

My cousin's dad and stepmother were familiar with the helmet. He had worn it for them one day in June 1969 while he was home on his last leave before departing for Vietnam. Uncle Louie, as an Army Air Forces veteran of World War II, was especially proud of his son, the newly minted warrant officer, and wanted to be sure to have a permanent record of the occasion. So Nicky put on his dress-blue uniform, with white shirt and bowtie, for picture-taking in the living room of Louie and Bert's rancher on King Road—the home he had moved to when he was eighteen. For one shot, Nicky put his arm around Bert as

175

both looked at the camera and smiled. For another, he faced his dad and earnestly shook his hand.

I can hear Uncle Louie saying, "Let's get one with your helmet on," and maybe Nicky winced at the idea because he'd look odd wearing the helmet with his dress blues instead of an unadorned flight suit. But he would do it anyway to please his dad. In the picture, he's sitting on the edge of a credenza, his visor's black shade pulled down so he appears bug-eyed, his arms at his sides and both hands fisted. In a photo I have framed and on my desk so I see it every day, he is at the table, chin in hand and looking away from the camera. It's supposed to be a happy time for him, at home with his folks, but his expression is grim and distant, sad and fearful, as if at the moment the shutter clicked, he could see his fate.

In a few weeks, he would be dead, one of 9,107 Americans whose deaths in Vietnam were classified as accidental.

He never wore the helmet in Vietnam; he was not alive long enough to be assigned to a company or to fly a helicopter there. The Army sent the headgear in its cloth cover with his other belongings, including a foot-long metal box in which he kept personal effects. They lay for many years in Louie and Bert's attic with Nicky's uniform and a large pile of other things he had accumulated in his year in the Army. After Louie died in 1996 and Bert followed him ten years later, the attic trove passed to Nicky's younger brother, L.B. He kept it in a rented storage shed in Coatesville, Pennsylvania, twenty miles down Route 30 from their boyhood home. One day he took me there and offered me everything.

Nicky's belongings are mine—laminated cards for preflight cockpit procedures, a foldout card on how to read coordinates and find direction by stars, a two-inch-thick operator's manual for the UH-1D helicopter, notes Nicky took in his classes at Forts Wolters and Rucker, a September 2, 1968, statement he gave the Army in which he swears that his only run-in with the law was a speeding offense in April that year on Route 30, for which he paid a $15 fine. There are ten wallet-size photos he took to Vietnam. One is of Louie and Bert. The rest show Terri Pezick, who would have been his wife. And there is Nicky's Army clothing—neatly pressed khaki shirts and pants, three pairs of shiny black boots and a pair of black dress shoes, a green service cap with black visor, olive-green jackets and even three pairs of socks.

I return to these objects over and over because holding them, reading them and looking at them bring Nicky alive. They were a part of him, things that touched him, things he used. They keep me connected to him and grounded with the purpose of telling his story.

It is a story that, for all of my efforts, will never be complete and has left me deeply frustrated and bitterly disappointed. I can't accept how callously his life and the lives of those who died with him were shoved aside, casualties of what I see as the Army's bureaucratic disinclination to examine its own mistakes. At the beginning of my inquiry, I had more faith in the military. I thought Nicky's whole story would be laid out clearly in the records, including the consequences of the deadly incident. Instead what I saw was confusion, chaos, ambiguity and the absence of anything that even approaches thoroughness. It's maddening. It disregards him, his comrades and all those who loved them.

A man who shared my assessment was a Vietnam-era Army criminal investigator not involved in Nicky's case or privy to its handling, but who knew the military's investigative process—how things were supposed to be done. "I can't fathom the degree of injustice that took place at the time and went totally undocumented by a competent higher authority," Carl Craig wrote to me. "What a travesty!"

The circumstances of Nicky's death were shocking and unusual, even in a war zone. A class on grenade safety for newly arrived troops had become the opposite, a death trap inside four walls. The explosion happened in the controlled environment of a classroom, in daylight on a highly trafficked landing zone that was the base for an entire light infantry brigade. Two warrant officers and a specialist died and a few dozen soldiers were wounded, some seriously. The commanding general of the Americal Division, the largest U.S. division in South Vietnam, called it the worst training accident that happened on his watch. In all of the Vietnam War, I could find only one other training tragedy that compared to it—the March 1967 explosion in the mine class outside Da Nang that killed those thirteen Marines.

And yet what passed for an investigation at LZ Bayonet strikes me as far less than diligent. Wayne Hesston, the instructor who lobbed his fragmentation grenade into the crowd of new arrivals, said he was not asked about it for weeks, not until after he had been returned to the States. Two warrant officers who were sitting with Nicky when the grenade detonated don't remember being interviewed about what they had seen and heard. The grenade is a mystery. Hesston remembered that with the one he had used in previous classes, he had removed its fuse and detonator. Somehow, when he brought his props from the cadre room into the classroom where Nicky, Billy Vachon and Tim Williams sat in front of him, he instead had a grenade that was ready to blow.

How had previous instructors handled the class? Anecdotally, I knew that the grenade-tossing routine had been used late in the previous year, 1968, more than six months before Hesston got the assignment. But I was never able to find anyone else who had taught the class. Robert C. Bacon, who as a lieutenant colonel was briefly in charge of the unit that provided orientation and training for new arrivals after Nicky's death, called it a risky and unnecessary attention-getting stunt, and told me there was a better way. "One of the best attention-getters was to say at the start of the class: 'Probably either you or the man sitting next to you will be killed or wounded during your tour. If you pay attention, it might not be you.'"

The Americal Division's commander, Lloyd B. Ramsey, had told me he was shocked by the severity of the accident. He remembered that there had been an investigation—he didn't remember by whom— and that someone had briefed him on the results: The explosion had been an accident and no one could determine how the instructor happened to have a live grenade. The Army apparently lacked any real interest in pursuing the case and didn't commit the facts to paper. If it had, we at least would know it had taken this ghastly loss of life seriously. At best, it appears to me, there was a breakdown of process.

It doesn't matter that Nicky's grief-stricken parents did not seek the full details of how he died. To them, the only thing that mattered was that he was gone. And yet the government, which has always been so proficient at training and sending its citizens off to war but so often has neglected them and their families afterward, let them down. A letter I got from Tim Williams' girlfriend reflected that failure. Tonie Vicario, who had written to me regularly from her home in Las Vegas about my efforts and the "sweet and polite" guy who wanted to marry her in 1969, said that while battling cancer she had made a point of speaking with veterans about Tim and his sacrifice. They told her that anyone who died on foreign soil in the country's service, no matter the circumstances, should be recognized by the government and their families consoled. Tonie clung to that until her death in 2013, believing that in Tim's case the Army should have done much more.

Some public recognition of Nicky, Billy and Tim has come from other quarters. A memorial plaque honoring "Patriots who died for their country" and naming Nicky and three other Great Valley High

School graduates killed in Vietnam stands in a display case at the school outside Malvern. Nicky's name is also on the Wall of Heroes in the old Chester County Courthouse, a list of the county's sons who died in Vietnam. In South Portland, Maine, Billy's flight school graduation photo hangs on a wall at the VFW post, and in 2002 the city councilors named a street after him. In Ohio, the American flag that draped Tim's casket was displayed in the field house at his alma mater, Rossford High School, for several years beginning in 1971, a gift from his family.

But we are left with the travesty of not having any official report on the record to examine now and in the future.

There is an objective reality for 10:15 a.m. July 10, 1969. I've often imagined myself being with Nicky on LZ Bayonet, seeing and hearing everything that transpired, second by second, until the critical moment and beyond. The reality I must accept, though, is that I will never know precisely what happened on the day fate turned against my cousin and the others.

From time to time, I've written to the Army to see if a report of investigation might have turned up since my early years of looking. But each time, most recently in 2015, I got the same response: We have nothing on this accident. In the fall of 2008, I spent a day at the National Archives in College Park, Maryland, with hands-on assistance from Vietnam records expert Rich Boylan. We found nothing more than brief mentions of the explosion in the daily journals of the 198th Light Infantry Brigade and the Americal Division.

Instructor Hesston said no action was taken against him—and there is nothing in his military records about the explosion. But in the angry phone message he left for me in 2013, he told me something new, a nugget of information that had never come out in our conversations before, that I never thought to ask him. It came after a pause.

In our first meetings, when Hesston mused about what happened in the orientation building and posed his idea of sabotage, he had told me several times that anyone could have gotten into the small cadre room where he kept his demonstration grenade and other props. Anyone could have gotten in there, he said, and tampered with his grenade or substituted a live one.

But now on the phone, after reminding me that "in war, stuff happens," he said without further comment: "The cadre room was locked up."

～

From the start of my efforts to learn about Nicky and what happened to him, I had assumed that every piece of information I gleaned would take me where I wanted to go. I would find out things about the explosion that would fit like pieces of an intricate puzzle. And when the last piece fell into place, I could look at the big picture with satisfaction. But my work did not end neatly, because some of the pieces are missing or don't fit. So I have had to settle for two parts instead of a whole.

One is resentment from seeing Nicky diminished by inattention, as were the others who died or, like Hesston, were injured and traumatized. But the other part has enriched me beyond measure: my discovery of who Nicky was. This cousin of mine who once pitched Little League baseball, hunted raccoons in the woods, and fussed over fast cars was every man yet, like all persons, unique. Along with the two others who died with him, he mattered in specific ways large and small. He was loved and cherished, grieved for and missed by concentric circles of people whose lives he had touched.

My earliest efforts had brought me into contact with Nicky's closest friend in the Army, Tony Viall. They had taken all their training together, gone to Vietnam at the same time and bled together as they crawled out of the fragment-riddled orientation building among the shrieks and cries of other men. On a week of vacation in the fall of 1996, I made the fifteen-hour drive from Allentown to meet Tony at his home in the small community of Ooltewah near Chattanooga, Tennessee. Over several days, he told me his story. And he remembered the name of another pilot who was on the bench with them when the grenade went off—Tom Sled.

I would find him, too.

Tom still suffers the effects of his wounds—his right leg had been gashed by fragments from mid-calf up to his thigh, the peroneal nerve shredded at the knee. "I still have drop foot and about thirty percent feeling in the leg below the knee and ten percent in the foot," he told me in 2004.

One afternoon in *The Morning Call* newsroom, a co-worker and I were talking about the explosion and he asked, "Are you going to go to Vietnam?" I seized on the idea immediately. Though it was unlikely I'd find answers about Nicky's death, at least I could get a sense of the place and follow the path he took just before he died. Some contacts I'd made put me in touch with a travel agency that specialized in helping Vietnam veterans return to their old battlegrounds. I made the trip in

May 1998, experiencing in those eleven days a land, climate and culture far different from what I've known in a life spent in Pennsylvania.

This Vietnam, with its peasants and rice paddies and steamy heat, brought me face to face with a regrettable past. At a dusty crossroads along the coast south of Da Nang, when I asked a man how he'd lost his leg, he pointed skyward and said, "American bombs." A woman who showed me around the My Lai massacre site said gravely, "We are not angry with all Americans, only the ones who did this." I couldn't help but feel sad and uncomfortable.

Tim Williams' sister, Jill Bell, shared similar feelings when she wrote to me: "I am sickened when I think back upon this shameful time in America's past. So many young men wasted, so many families torn apart—all pawns of a political system that were drawn into a civil war on the opposite side of the world. I am still bitter, as well as very distrustful and apathetic toward anything political."

My trip to Vietnam came at the end of a six-month unpaid leave of absence I'd taken to work on the project. Earlier in my leave, Mary and I had driven to Portland, Maine, to meet Billy Vachon's family and see how they had coped with their loss. In the summer of 1969, Billy came home from Vietnam on the same flight as Nicky and was buried July 26 in South Portland's Calvary Cemetery. The Vachons commemorate his life on every anniversary of his death by having a Mass celebrated for him. They pay to run his picture in the *Portland Press Herald* with the message: "We will always remember and love you."

On his parents' kitchen wall hung a painting of Christ. "He's my friend," said Billy's mother, Louise. Every room had religious pictures or figurines—symbols of the Vachons' Catholic faith. Photos of Billy, his brothers and sisters and their children crowded the wall at the foot of the stairs.

Billy's father, who had fought the Nazis in an antiaircraft artillery battalion and was called Bill, remembered that after Germany fell and he was going through training in France to prepare him for the fight against Japan, an Army instructor tossed a German potato-masher grenade at him and several dozen other soldiers. They knew it was a dud and didn't scatter but instead "laughed like hell." The joke came back to Bill Vachon twenty-four years later, when a similar training prank on the other side of the world killed his first-born son.

Louise mentioned that a box of Billy's things was upstairs. She wasn't sure what was in it because she hadn't gone through it since his death.

"I should get it," she said, but she didn't go.

Mary and I slept upstairs in Billy's old room, an intimacy that brought him closer, made him more real. The next day, Louise retrieved the cardboard box of her son's cards, letters and Army telegrams that she had hesitated to bring downstairs and open. She pulled out a note his young daughter scrawled after her dad died.

> Dear father, I wish you never die. Because it seems if I did't know you that well. I wish we could be to gether agin. I hope you would come back again. And I hope you are being good up their. I wish you did't die. So we could be to gether agin. By-by Daddy. Love, your daughter Tina Vachon

We met Billy's high school friend and fellow athlete Larry Hogan, who fought in Vietnam and was like others I would get to know who came home from the war forever changed. Larry said that in his year with the 199th Light Infantry Brigade, he had become so accustomed to fear, savagery and the adrenaline high of a firefight that he worried about how he might behave back in "the World." When his time was up, he signed on for five more months. He showed me a poem he wrote in 1969 in which he begged: *Someone please try and save us. Someone love us; we are scared.*

I took some additional time off in 2001 to work on Nicky's story. That was when Mary and I flew to California to see Hesston. All along, I juggled the responsibilities and diversions of job and family as I pieced together my cousin's short life. There were long periods that I put the story aside and did nothing with it, didn't even look at it, unsure that I would ever be able to tell it in a meaningful way. At one point I had completed a manuscript and tried to interest literary agents and publishers but got no nibbles.

In the meantime, my pursuit of Nicky's story steered me toward work at the newspaper I never would have sought otherwise. I'm not a veteran and previously had no particular interest in the military, though my dad, an ex-sailor, had tried to nudge me into the Coast Guard. But my interviews for Nicky's story have drawn me to people who've served the country. I heard personal accounts of the Vietnam War, and in talking to my uncles and an aunt about Nicky, was surprised to learn that they had served in World War II in far-flung places and had extraordinary experiences—tales worth telling and recording for posterity.

In 1999, I began interviewing veterans in the greater Lehigh Valley for a series called "War Stories: In Their Own Words." These were interviews I recorded, transcribed and shaped into a narrative. I wasn't writing *about* the veterans; I was presenting their personal accounts as though they were speaking to the reader—a print version of the Library of Congress' audiovisual Veterans History Project. Since then, I've done about 110 interviews with veterans from the World War I era, World War II, the Cold War, and the Korean and Vietnam wars. In 2011, *The Morning Call* published a collection of my stories in a book, *War Stories in Their Own Words.* So, as a result of my interest in Nicky, I have helped to preserve valuable accounts for future generations.

That work helped me stay focused as I wrestled with disappointments in advancing Nicky's story. But all along, I refused to let go of him. I'd reach back to that day when I was a clueless fifteen-year-old at his funeral, feeling the weight of grief on those around me, or I'd take a few moments to gaze at his picture on my desk. He was in my heart. His life and the mystery of what happened to him never have stopped tugging at me.

My years of searching granted me some success and peace of mind. I was able to build a portrait of Nicky—a working-class kid trying to find his way in the tumultuous late 1960s, making wrong turns but ultimately finding purpose in leaving behind all he knew to fly above the jungles, mountains and rice paddies of an alien land. I came to feel as though I had gone back to that era and walked side by side with him, sharing a common American experience but at the same time reaching the realization that he and the others lost were irreplaceable. Friends and relatives, sharing their remembrances and mementos, have brought me closer to him than I ever could have imagined.

I wish he were here. I bet we'd be friends.

Though Nicky lived only to age twenty, he touched other lives in enduring ways. Mary Anne Wallace, who was his friend but wanted to be more than that, still carries a picture of Nicky in her wallet and visits his grave. "He was a good friend," she said. "He was the first guy I ever really cared about." The St. Christopher medal she had given him for protection had been sent back from Vietnam, one of the two, unspecified religious medals listed on the Record of Personal Property in Nicky's casualty file. Mary Anne was gratified to learn that he had taken it with him. When I returned it to her, she hugged me.

One of the most extraordinary friendships Nicky made was with his buddy Tony's younger sister, Debbie. She had grown fond of Nicky when he spent weekends at Tony's home in Rossville, Georgia, while the two young men were in flight school at Fort Rucker, Alabama. She would lend her room to him, and he kept his clothes in a closet she rarely used.

"I remember the day, after hearing of Nick's death, that I opened the seldom-used closet and found his Army garment bag, his name printed on it, hanging there," Debbie wrote to me in 1996 from her home in Knoxville, Tennessee. "We could not bring ourselves to dispose of it, and so it hung in my closet long afterward.... We felt that the disposal of Nick's property made his death too real for us."

She had known Nicky only briefly, and not well, but enough to believe he was "honorable in complying with the requirement of his country—a man who did not run, nor buy or fake his way out, though in retrospect we wish he had."

She named her son after him.

Her brother Tony continues to grieve over what happened that July morning in Vietnam and still misses his friend. "I don't think I ever told you this before," he said over the phone one day when I was just about to hang up. "Remember I said Nick always used to sing 'My Girl'? Well, I listen to an oldies station in my house. Every time I hear that song on the radio, I turn it off.

"I automatically think of Nick, after all these years."

Then there is Nicky's girl. Terri Pezick remembers that before he left for the war, Nicky urged her to live happily if he did not return. Despite the shock and heartbreak of his loss, she did her best to obey. She was married, had two sons and a full life when I met her in 1996, more than a quarter century after Nicky's death. The days she had spent with him were so far behind her, they must have seemed like a shadow of a life other than her own. Yet I could tell that Nicky still touched her. Once, I held up the videocassette that had the movie images of her and Nicky from June 1969 and asked if she'd like to watch the tape. She looked into my eyes but seemed to drift away as the seconds passed.

"No," she finally said, "I don't think I could do that."

Late one night in 1998, as I was at my computer in my attic office writing about my cousin, I felt a presence.

Nicky stood behind me, smiling and looking over my left shoulder at the screen. I felt him say "hi." Though he spoke no words, he conveyed this kindly message: I'm embarrassed at the attention you're giving me, but grateful.

The presence vanished, and I got back to work.

The Author's Interviews

Anderson, Ralph, served in 723rd Maintenance Company, Americal Division, Chu Lai, and was friend of explosion survivor Bob Beck. April 18, 20, 22, 23, 28 and 30, 2013, by email; May 8 and 9, 2013, by email; September 19 and 23, 2013, by email

Appy, Christian G., professor of history, University of Massachusetts at Amherst; on communication between Army and Marine Corps in Vietnam. March 11, 2014, by email

Atwater, Jack, a Vietnam veteran who was director of U.S. Army Ordnance Museum at Aberdeen Proving Ground, Maryland; on workings of a grenade. February 24, 1998, by phone

Bacon, Robert C., was commandant, 23rd Adjutant General Replacement Company, Chu Lai, and later commander of 3rd Battalion, 21st Infantry, 196th Light Infantry Brigade. Sept. 24, 1996, by phone; July 22, 1998, by phone; Sept. 14, 1998, by letter; September 28, 1999, by letter; March 18, 2001, by phone; April 21, 2014, by phone

Bagley, Kevin L., a retired Army lieutenant colonel, on flight training; was Vietnam War helicopter pilot and Fort Wolters instructor. May 20, 23 and 28, 1997, by email; July 15, 1997, by email; April 21, 1998, by email; July 8, 1998, by email; February 7, 2001, by email; June 26 and 28, 2001, by email; October 7, 10 and 21, 2002, by email; November 1 and 5, 2002, by email; December 4 and 5, 2002, by email; February 13, 2003, by email; September 12, 16 and 17, 2004, by email; October 13 and 16, 2004, by email; November 15, 2004, by email; November 30, 2006, by email

Baker, Larry, on care at 312th Evacuation Hospital, Chu Lai; was with 17th Cavalry, 198th Light Infantry Brigade. March 21, 2000, by letter

Baranzano, Larry, one of Nicky's hometown friends. February 6, 1998, in person

Bartlett, Stanley L., was headquarters company commander, 16th Combat Aviation Group, Chu Lai. February 8, 1998, by phone

Beam, Bill, Nicky's cousin. August 1, 2001, by phone

Beam, Mike, Nicky's cousin. January 26, 1997, in person

Bedics, Lynn (O'Malley), was an Army nurse at 312th Evacuation Hospital, Chu Lai, later the 91st Evacuation Hospital. April 6, 1999, in person; September

11, 2000, by phone; March 9 and 10, 2006, by email; September 19, 2008, by email; October 17, 2008, by email; May 30, 2010, by email; February 5, 2011, by email; February 8, 12 and 14, 2012, by email; March 6 and 12, 2012, by email; April 4 and 22, 2012, by email; July 26 and 29, 2012, by email; March 27, 2013, by email; April 16 and 17, 2013, by email; March 25 and 27, 2015, by email; March 29, 2015, in person; April 1, 8, 10 and 14, 2015, by email; July 4, 5, 9, 12, 23 and 24, 2015, by email; August 7 and 14, 2015, by email

Bell, Jill (Williams), Tim Williams' sister. June 7, 1998, by email; November 8, 2001, by email

Billington, Dr. Bradley I., worked at 27th Surgical Hospital, Chu Lai. October 7, 1998, by phone

Bleier, Rocky, Americal Division veteran and former Pittsburgh Steelers running back. April 15, 1998, by phone

Boehmler, Charley, one of Nicky's best friends. Nov. 27, 1996, in person

Boehmler, Kathryn, Charley's wife. April 18, 2001, in person; April 28, 2002, by email

Bowen, G. "Bud," was with 54th Medical Detachment (Helicopter Ambulance), Chu Lai. January 31, 2011, by email; February 1 and 14, 2011, by email

Bradshaw, Terry, was an Americal Combat Center truck driver, Chu Lai. February 7 and 10, 1998, by email; April 7, 9 and 11, 2014, by email

Caldara, Barbara, was an Army nurse at 312th Evacuation Hospital, Chu Lai, later the 91st Evacuation Hospital; tended to Billy Vachon. April 28, 1999, by letter; May 23, 1999, by letter; Aug. 3, 1999, by letter; Aug. 24, 2000, by email; February 9, 2001, in person; January 6, 2007, by letter; April 8, 2013, by email

Canady, Edward R., was commander, 23rd Adjutant General Replacement Company, Chu Lai. November 11, 1996, by phone; April 21, 2014, by phone; July 8, 2015, by phone

Cassell, Ernie, was an Americal Combat Center truck driver, Chu Lai. January 14, 2001, by phone; September 28, 2002, by email

Clarkson, Debbie (Viall), Tony Viall's sister. November 9, 1996, by letter

Clemons, Joseph G., Jr., was commander of Americal Support Command, Chu Lai, and 198th Light Infantry Brigade. July 30, 2013, by phone

Cockerham, William E., Sr., former police chief of Malvern. May 18, 1997, in person

Cohn, Frank, was a lieutenant colonel in Army's Provost Marshal Office at Long Binh, Vietnam; on investigative process. April 16, 18, 19, 21 and 22, 1999, by email; May 11, 1999, by email

Conroy, John, Vietnam veteran. March 25, 1999, by email

Craig, Carl, former Army Criminal Investigation Division liaison officer in Tokyo 1966–69. July 5, 6, 10, 11, and 30, 2013, by email; July 7, 2013, by phone; Aug. 13, 2013, by email; October 11, 14 and 15, 2013, by email; March 27 and 29, 2014, by email.

Curtis, Clyde, was an Americal Combat Center instructor. September 2, 1998, by phone

D'Addario, Raymond E., former chief of Criminal Investigation Division, 25th Infantry Division. September 23, 24 and 28, 1999, by email; October 1, 1999, by email

Dieli, James B., an Americal Division veteran. October 28, 2013, by email

Dougherty, Lorraine (Pusey), Nicky's half-sister. July 29, 2001, by phone

Ducceschi, Nikki (Venditta), Nicky's cousin. January 15, 1998, in person

Elliott, Stanley, was in 1st Battalion, 52nd Infantry, 198th Light Infantry Brigade, LZ Bayonet. September 12, 1998, by letter; September 27, 1998, by phone; October 19, 1998, by letter; March 22, 1999, by letter; February 12, 2000, by letter; April 6 and 24, 2000, by email; May 1, 2000, by email; September 11, 2000, by email; November 6, 2000, by email; December 24, 2001, by email; February 19, 2003, by letter; April 25, 2003, by letter; July 16, 2003, by letter; June 28, 2004, by email; December 22, 2004, by letter; June 24, 2005, by letter; June 24, 2008, by email; December 8, 2008, by email; March 20, 2010, by email; June 17, 2011, by email; August 9, 2015, by phone; August 9 and 12, 2015, by email

Elmer, Victor J., was Army officer who escorted Nicky's body home from Dover Air Force Base, Delaware. February 1, 1999, by phone

Eversole, Tad, was in 1st Battalion, 14th Artillery, 198th Light Infantry Brigade, LZ Bayonet. September 24, 1998, by phone; October 16, 1998, by letter; December 17, 1998, by letter; November 19, 1999, by letter; April 21, 2015, by email; July 22, 2015, by email

Feasel, Larry S., was Army helicopter pilot in Americal Combat Center training with Nicky, flew with 176th Assault Helicopter Company. May 13, 14, 15 and 16, 1997, by email; June 26, 1997, by email; July 21, 1997, by email; February 1, 1998, by phone

Feher, Andre C.R., an Army Criminal Investigation Division investigator of the My Lai massacre, on procedure. April 23, 1999, by letter; October 31, 1999, by letter

Forrester, Denny, one of Nicky's hometown friends. January 15, 1998, in person; January 18, 1998, by email

Forrester, Gerry and Barney, Nicky's friends and parents of Denny, Steve, Patricia. January 15, 1998, in person

Forrester, Patricia, one of Nicky's hometown friends. January 7, 1998, in person; January 10, 1998, by email; January 30, 1999, by email

Forrester, Steve, one of Nicky's hometown friends and a Vietnam veteran. January 22 and 30, 1998, by phone

Frame, Thomas, was explosion survivor Bob Beck's hometown friend and later his brother-in-law. June 1, 2013, by phone

Fritz, Kenneth H., served with 176th Assault Helicopter Company, Chu Lai. June 5, 1997, by email

Geserick, Edward W., Jr., was an Americal Combat Center cook and artist. March 14, 2001, by phone

Goodhart, Linda, was an Army nurse at 312th Evacuation Hospital, Chu Lai. May 15, 2003, by phone

Grande, Al, Serious Incident Reports officer in the Office of the Provost Marshal General of the Army in 1972. April 18 and 20, 1999, by email

Gray, Carl W., was a buck sergeant with 1st Battalion, 14th Artillery, 198th Light Infantry Brigade, LZ Bayonet. March 3, 8 and 10, 1998, by email; June 7, 1998, by email; June 2, 1999, by email; February 17, 2002, by email; October 17, 2008, by email; November 13, 2008, by email

Gray, Joe, Jr., Nicky's stepbrother. Nov. 27, 1996, in person; October 4, 2001, in person; October 14, 15, 21, 26 and 28, 2001, by email; October 25 and 30, 2002, by email; December 28, 2006, by email; May 1 and 2 and 13, 2011, by email; July 10, 2011, by email; September 8 and 20, 2011, by email; September 18, 2011, in person; October 20 and 21, 2012; June 21 and 23, 2014, by email; July 4, 2015, by email

Gregg, Jackie Raushi, one of Nicky's hometown friends. March 11, 1998, by phone

Gross, Dr. Alton F., operated on Nicky at 27th Surgical Hospital, Chu Lai. June 4, 1998, by phone

Gunn, L. Ray, associate professor of history, University of Utah, on communication between Army and Marine Corps in Vietnam. March 18, 2014, by email

Harp, Ray E., of U.S. Army Training and Doctrine Command, Office of the Chief of Public Affairs, on grenade training. April 30, 2014, by email

Hayden, Jess, from Malvern, was a Marine machine-gunner in Vietnam, lost left leg below knee to a grenade in December 1969, knew Nicky and his father, Louie. April 2, 1998, in person

Hazlett, Myrtle (Gable), Nicky's aunt. November 10, 1997, by letter; July 29, 2001, by phone

Hesston, Wayne, pseudonym for the grenade instructor at LZ Bayonet. November 13 and 17, 2001, in person; November 23, 2001, by phone; December 1, 2001, by phone; January 20, 2002, phone message; February 2, 2002, phone message; August 12, 2002, by phone; September 6, 2003, phone message; January 18, 2005, by phone; December 13, 2005, by phone; September 15, 2012, by phone; July 20, 2013, phone message

Hibberd, Josiah, was Nicky's teacher and principal at Malvern Public School. February 2, 1998, by phone; March 20, 1998, in person

Hogan, Larry, Billy Vachon's hometown friend. January 5, 1998, in person; September 22, 2004, by phone

Hoggard, Dennis, Sr., was stationed at LZ Bayonet with 198th Light Infantry Brigade. December 29, 1998, by phone

Holdridge, David, drove truck from American Combat Center, Chu Lai, to LZ Bayonet. July 24, 1999, by phone; September 21, 1999, by phone

Hovde, Jon, Vietnam War amputee and author of *Left for Dead: A Second Life After Vietnam*. January 7, 8 and 10, 1999, by email

Hunt, Dr. Robert S., served at 27th Surgical Hospital, Chu Lai. Week of August 16, 1997, by phone; July 8, 1998, by letter

Irwin, Thomas J., a lieutenant colonel, was provost marshal of 23rd MP Company, Chu Lai, until July 7, 1969; on investigative process. March 20 and 22, 1999, by email; April 11, 12 and 29, 1999, by email

James, Darryl, was a pilot in Division Artillery Aviation Section, Americal Division. August 4, 1997, by email; June 23 and 30, 1998, by email

Johnson, Hank, Vietnam veteran and author of *Winged Sabers: The Air Cavalry in Vietnam*. January 7, 1998, by email

Joye, Bobbie (Stiteler), one of Nicky's hometown friends and best friend of his fiancée, Terri Pezick. February 2, 1998, by phone

Kendzezeski, Mike, served with 11th Light Infantry Brigade, American Division. July 22, 26 and 28, 1999, by email

Kenworthy, Karen (Armstrong), one of Nicky's Great Valley High School classmates. January 29, 1998, by phone; October 13, 2001, by phone

Kerns, Thomas C., was operations officer at American Combat Center, Chu Lai. May 11, 1997, by phone; April 21, 2014, by phone; June 16, 2015, by phone; July 12, 2015, by email

Kilgallon, the Rev. John, was Catholic priest for Nicky's graveside service. May 14, 2011, by phone

King, Joseph C., one of Nicky's hometown friends. August 9, 1998, in person; October 6, 2001, by phone

Kinman, Dr. Philip B., was battalion surgeon in 196th Light Infantry Brigade, July 3, 1997, by phone

Knox, Dave, operating room technician at 27th Surgical Hospital, Chu Lai. March 7, 9, 10 and 29, 1998, by email; June 24 and 25, 1998, by email; July 19, 27 and 29, 1998, by email; December 24, 1998, by email

Kocmalski, Pete, operating room technician at 312th Evacuation Hospital, Chu Lai. July 23, 24, 25, 26, 27, 28, 30 and 31, 2012, by email; May 27, 2013, by email

Kralich, Joe "Doc," an American Division veteran. May 5, 1997, by email; September 3 and 24, 1997, by email

Lee, Miriam A. (Vachon), Billy Vachon's widow. July 24, 1996, by phone; January 6, 1998, by phone; February 21 and 22, 1998, in person; October 19, 1998, by email; March 15 and 16, 1999, by email; April 22, 1999, by email; March 14, 2000, by email; May 4, 2000, by email; August 14 and 21, 2000, by email; November 21, 2001, by email; May 28, 2003, by email; December 10, 2003, by email; August 18, 20, 27 and 30, 2004, by email; September 8, 2004, by email; October 16, 21 and 22, 2004, by email; December 17 and 21, 2004, by email

Lindholm, Steve, one of Billy Vachon's flight school friends and a Vietnam veteran of 1st Cavalry Division (Airmobile). August 8, 11, 12 and 19, 1997, by email; September 5 and 22, 1997, by email

Loffgren, Max, served with 1st Battalion, 52nd Infantry, 198th Light Infantry Brigade, LZ Bayonet. December 28, 1997, by email; January 1, 1998, by email; June 24, 1998, by email

Luke, Lance, an American Division veteran. December 3 and 5, 1999, by email; December 12, 1999, by phone

Marble, Dr. Sanders, historian, Office of Medical History, Office of the Surgeon General, U.S. Army, on Vietnam War casualty notification. September 22, 2008, by email

Marrash, Dr. Samir, Nicky's physician at 312th Evacuation Hospital, Chu Lai. June 1 and 13, 1999, by phone

McLaughlin, Frank "Max," was a mess hall baker at American Combat Center, Chu Lai. March 14, 2001, by phone

Mullen, Peg, mother of Nicky's fiancée, Terri Pezick. January 27, 1998, by phone; February 2, 2001, in person with her daughter and Nicky's stepmother, Bertha Venditti

Nerone, Francis A., served as assistant chief of staff for intelligence at American Division headquarters, Chu Lai. July 20, 2013, by letter

Noller, Gary L., of the Americal Division Veterans Association. January 13, 2000, by email; April 16, 2000, by letter; September 29, 2000, by email; November 3, 2005, by email

Parmantier, Gerald J., was a medic with 198th Light Infantry Brigade rear aid station at LZ Bayonet. March 3 and 14, 1999, by email

Peterson, Bob, one of Nicky's hometown friends. April 4, 2001, by phone

Pettine, Patty (Venditta) and Jimmy, Nicky's aunt and uncle in Malvern. January 20, 1998, in person

Pilkinton, Samuel T. III, served with 26th Combat Engineer Battalion, Americal Division. December 22, 1997, by email; December 28, 1997, by phone; April 30, 1998, by phone; December 18, 1998, by letter; April 18, 1999, by phone; September 17, 1999, by phone; October 9, 2005, by phone; August 13, 2012, by letter; September 4, 2012, by letter

Powell, Colin L., was deputy assistant chief of staff for operations at Americal Division headquarters, Chu Lai. March 20, 2005, by letter

Pusey, Sally (Venditti) and John, Nicky's mother and stepfather. March 3, 1996, both, in person; May 30, 1996, Sally, by letter; January 27, 1998, Sally, in person; November 17, 1998, both, in person; January 22, 1999, Sally, by letter; February 12, 2000, Sally, by letter; June 8, 2000, Sally, by letter; August 23, 2000, Sally, by letter; October 10, 2000, Sally, by letter; March 9, 2001, both, in person; March 10, 2001, Sally, by phone; July 13, 2001, Sally, in person; October 3, 2001, John, in person; October 13, 2005, John, by phone

Ramsey, Lloyd B., retired Army major general who was commander of the Americal Division at Chu Lai. November 18, 1997, by phone; January 8, 1998, by phone; July 5, 1998, by letter; August 9, 1998, by letter; September 20, 1998, by letter; November 20, 1998, in person; October 10, 1999, by letter; December 7, 1999, by letter

Randolph, Nancy, was an Army nurse at 312th Evacuation Hospital, Chu Lai. March 14, 1999, by phone

Reed, John S., Iraq War veteran and associate professor of history, University of Utah; on communication between Army and Marine Corps in Vietnam. March 24 and 26, 2014, by email

Rishel, Charles "Art," served in 1st Battalion, 46th Infantry, Americal Division. March 29, 2001, by phone

Robinson, Emmet, Sally and John Pusey's next-door neighbor. October 24, 2001, by phone; October 25, 2001, by letter

Schwarzkopf, H. Norman, retired Army general, commanded 1st Battalion, 6th Infantry in Vietnam and led coalition forces in Persian Gulf War; January 26, 1998, by letter

Shain, Dr. Tin Zar, psychiatrist at Sacramento VA Medical Center, March 13, 2001, by phone.

Shkurti, William J., Vietnam veteran, author of *Soldiering On in a Dying War* and adjunct professor of public affairs, Ohio State University; on communication between Army and Marine Corps in Vietnam. March 13 and 14, 2014, by email

Short, Robert G., served in 1st Battalion, 46th Infantry and was an Americal Combat Center instructor in 1970; November 9 and 10, 1997, by email; January14,

2001, by email; January 16 and 31, 2001, in person; February 5, 10 and 12, 2001, by email; February 16, 2001, by letter; March 21, 2001, by email; April 16, 2001, by email; June 11, 2001, by email; January 26, 2002, by email; February 7, 19 and 26, 2002, by email; July 15, 2002, by email; November 21, 2002, by email; February 13, 23 and 25, 2003, by email; March 3 and 5, 2003, by email; April 24, 2003, by email; May 1, 2003, by email; June 23, 2003, by email; February 5, 2004, by email; September 12 and 16, 2004, by email; November 18 and 28, 2004, by email; March 6, 2005, by email; October 13, 2005, by email; November 1, 2005, by phone; November 22, 2006, by email; October 29, 2007, by email; November 14, 2007, by email; April 29, 2008, by letter; September 8, 2008, by letter; December 3 and 30, 2008, by email; June 15, 2009, by email; March 28, 2014, by email; May 8, 2014, by email; June 11, 2014, by email

Sled, Thomas, helicopter pilot in 16th Combat Aviation Group, Americal Division, who was wounded in explosion that killed Nicky. November 8, 1996, by phone; November 12, 1996, by email; January 5 and 25, 1997, by email; May 14, 15 and 16, 1997, by email; July 29, 1997, by email; August 12, 1997, by email; September 29, 1997, by email; October 20, 1997, by email; November 10, 1997, by email; January 29, 1998, by phone; February 9, 1998, by letter; March 9 and 14, 1999, by email; March 6 and 19, 2000, by email; May 5, 2000, by email; July 15, 2002, by email; October 23, 2002, by email; March 11, 12, 13, 14 and 27, 2003, by email; September 13, 14, 16, 17, 18, 20, 22 and 27, 2004, by email; February 22, 2005, by email; July 15, 2009, by email

Smith, Edwin A. "Skip," Jr., was Army helicopter pilot and one of Tony Viall's and Nicky's friends and flight school classmates. March 28, 1999, by phone

Spivack, Dr. Alan R., was chief of medicine at 27th Surgical Hospital, Chu Lai. September 23, 1997, by phone; September 25, 1997, by email; May 5, 1998, by email

Sproehnle, Gerald A., was a medic at 312th Evacuation Hospital, Chu Lai. September 7, 1998, by phone; September 12, 1998, by letter

Stewart, Terri (Pezick), Nicky's fiancée. March 24, 1996, in person; February 6, 1998, in person; October 18, 2000, by phone; January 22, 2001, in person; February 2, 2001, in person with her mother, Peg Mullen, and Nicky's stepmother, Bertha Venditti; July 27, 2001, by phone; October 8, 11 and 15, 2013, by email

Stone, Steve W., was Army helicopter pilot in Americal Combat Center training with Nicky, flew with 176th Assault Helicopter Company, Chu Lai. December 17, 1997, by phone; February 1, 1998, by phone; March 17, 1998, by letter; May 3, 1998, by phone; March 31, 2014, by letter; April 7, 2014, by phone; April 7, 22 and 24, 2014, by email

Stuhr, Donald R., served with 1st Battalion, 14th Artillery, 198th Light Infantry Brigade at LZ Bayonet. September 13, 1998, by phone; soon afterward by letter

Teders, Steven J., clinical psychologist at Allentown Veterans Affairs Outpatient Clinic, September 14, 2000, by phone.

Thomas, Charles E., Air Force veteran of Korean War; worked at gas station with Nicky and went street-racing with him. February 2, 1998, by phone; February 8, 1998, in person

Tritt, William J., Jr., one of Nicky's Army flight school classmates, served in Vietnam with 227th Assault Helicopter Battalion, 1st Cavalry Division (Airmobile). April 16, 1998, by phone

Underhill, George R., was commandant of 23rd Adjutant Replacement Company, Chu Lai. March 21, 2001, by phone; April 6, 2001, by letter; March 1, 2015, by email

Vachon, Louise and Wilbur J. "Bill," Jr., Billy Vachon's parents. August 11, 1996, by phone; February 20–22, 1998, in person; July 10, 1998, by letter; July 18–19, 1999, in person with Louise; August 12, 1999, by letter from Louise; January 10, 2001, by phone with Louise; November 5, 2001, by phone with Louise; December 9, 2001, by letter from Louise; June 14, 2003, by phone with Louise; October 2, 2003, by letter from Louise; September 8 and 21, 2004, by phone with Louise; August 15, 2006, by letter from Louise; December 1, 2007, by letter from Louise; September 20, 2008, by letter from Louise

Venditta, Elizabeth H., my mother. January 28, 1998, and on many other occasions about Nicky and our family

Venditta, Florence "Dutchy" and Frank, my aunt and uncle. January 15 and 19, 1998, in person; January 28, 1998, in person with Frank, my mother, and Patty and Jimmy Pettine; November 18, 1998, in person with Frank; August 8, 2001, in person with Frank

Venditta, Nancy, my cousin, daughter of my Uncle Sam and Aunt Ruth. July 13, 2001, in person

Venditta, William H., my older brother, on Nicky's funeral. April 22, 2011, by email

Venditti, Bertha and Louis C., Nicky's stepmother and father. December 8, 1995, in person; December 22, 1997, in person with Bert; January 19, 1998, in person with Bert and her son, Joe Gray, Jr.; January 28, 1998, in person with Bert; February 17, 1998, in person with Bert; June 27, 1998, in person with Bert and Joe; November 18, 1998, in person with Bert; February 2, 2001, in person with Bert, Peg Mullen and her daughter, Terri (Pezick) Stewart; April 11, 2001, in person with Bert; July 19, 2001, in person with Bert

Venditti, Harold "L.B.," Nicky's brother. March 9, 2001, in person; October 3, 2001, in person

Viall, Jerald A. "Tony," Nicky's best friend in the Army, went through boot camp and helicopter flight school with him; wounded in explosion that killed Nicky. February 26, 1996, by phone; March 8, 1996, by phone; October 24, 1996, in person; January 17, 1999, by phone; April 22, 1999, by letter; January 16 and 21, 2001, by phone; December 1, 2001, by phone; July 16, 2013, by letter; March 8, 2015, by phone; April 14, 2015, by phone

Viall, Jewell, Tony's mother. February 25, 1996, by phone; March 5, 1996, by phone; October 22, 1996, in person; October 6, 2002, by phone

Vicario, Tonie, Tim Williams' girlfriend. May 8, 1998, by phone; October 10, 2007, by letter; November 10, 2007, by letter; January 5, 2010, by letter

Vo Ngoc, Than, was a patient at 312th Evacuation Hospital, Chu Lai. April 8, 10 and 30, 2014, by email; May 5, 2014, by email

Wallace, Bonnie (Pusey), Nicky's stepsister. January 27, 1998, by phone

Wallace, George, one of Nicky's Malvern classmates, played baseball with him. June 5, 1996, by phone

Wallace, Mary Anne, one of Nicky's close friends. February 25, 1996, in person; August 15, 2001, by letter; November 21, 2001, by email

Welsh, David P., a retired Army chief warrant officer five, on contract status of warrant officer candidates. October 27 and 29, 2002, by email

Whittington, Jere O., was commander of 198th Light Infantry Brigade, LZ Bayonet. February 9, 1998, by phone

Williams, Adeline, Tim Williams' stepmother. April 14, 1998, by phone; December 15, 2000, by letter

Williams, Lila and Gary, Tim Williams' sister-in-law and brother. March 9, 1998, by phone with Gary; April 18, 1998, by phone with Gary; August 13, 1998, by letter from Gary; November 7, 2001, by phone with Gary; November 13, 2001, by email from Gary; November 10, 2002, by letter from Lila; December 7, 2002, by email from both; April 27 and 28, 2003, by email from Gary; May 2 and 13, 2003, by email from Gary; August 24, 2003, by email from Lila; October 22 and 25, 2004, by email from Gary; November 1, 2004, by email from Lila; December 18, 2004, by letter from Lila; March 5, 2008, by email from Lila; December 29, 2008, by email from Lila

Williams, Robert L. "Butch," Jr. Tim Williams' brother. March 6, 1998, by phone; May 15, 2003, by phone

Wright, Patricia, Joe Gray's stepsister and one of Nicky's friends. March 4, 1998, in person; October 19 and 20, 2001, by phone; May 1, 2010, by email; April 28, 2011, by instant message

Zaza, Robert N., was Army's chief Criminal Investigation Division investigator of My Lai massacre; on investigative process. April 22, 1999, by phone; September 9, 1999, by postcard

Note: Two Americal Division veterans I interviewed and who are quoted or paraphrased in this book asked not to be named. They are not on this list. Also, the Americal veteran who revealed the grenade instructor's name is not listed here.

Selected Reading

Bigler, Philip. *Hostile Fire: The Life & Death of First Lieutenant Sharon Lane.* Arlington, Va.: Vandamere Press, 1996.

Bryan, C.D.B. *Friendly Fire.* New York: G.P. Putnam's Sons, 1976.

Burkett, B.G., and Glenna Whitley. *Stolen Valor: How the Vietnam Generation Was Robbed of Its Heroes and Its History.* Dallas: Verity Press, 1998.

Caputo, Philip. *A Rumor of War.* New York: Holt, Rinehart & Winston, 1977.

Edelman, Bernard, ed. for The New York Vietnam Veterans Memorial Commission. *Dear America: Letters Home from Vietnam.* New York: W.W. Norton & Company, 1985.

Herr, Michael. *Dispatches.* New York: Alfred A. Knopf, 1977.

Karnow, Stanley. *Vietnam: A History.* New York: The Viking Press, 1983.

Kovic, Ron. *Born on the Fourth of July.* New York: McGraw-Hill, 1976.

Krakauer, Jon. *Where Men Win Glory: The Odyssey of Pat Tillman.* New York: Doubleday, 2009.

Lepre, George. *Fragging: Why U.S. Soldiers Assaulted Their Officers in Vietnam.* Lubbock: Texas Tech University Press, 2011.

Marlantes, Karl. *What It Is Like to Go to War.* New York: Atlantic Monthly Press, 2011.

Moore, Lt. Gen. Harold (retired), and Joseph L. Galloway. *We Were Soldiers Once...and Young: Ia Drang—The Battle That Changed the War in Vietnam.* New York: Random House, 1992.

Mullen, Peg. *Unfriendly Fire: A Mother's Memoir.* Iowa City: University of Iowa Press, 1995.

O'Brien, Tim. *The Things They Carried.* Boston: Houghton Mifflin, 1990.

Ramsey, Maj. Gen. Lloyd B. (retired). *Maj. Gen. Lloyd B. Ramsey, U.S. Army Retired: A Memoir.* Albany, KY: Clinton County Historical Society, 2006.

Schwarzkopf, Gen. H. Norman (retired), with Peter Petre. *It Doesn't Take a Hero.* New York: Bantam Books, 1992.

Sheehan, Neil. *A Bright Shining Lie: John Paul Vann and America in Vietnam.* New York: Vintage Books, 1989.

Teglia, Kenneth F. *Lamb in a Jungle: Conscience and Consequence in the Vietnam War.* Medina, OH: War Journal Publishing, 2012.

Turse, Nick. *Kill Anything That Moves: The Real American War in Vietnam.* New York: Metropolitan/Holt, 2013.

Index

www.ingramcontent.com/pod-product-compliance
Lightning Source LLC
Chambersburg PA
CBHW021142090426
42740CB00008B/895